GEOLOGY AND ENVIRONMENT IN BRITAIN AND IRELAND

CONTENTS

CONTENTS

GEOLOGY AND ENVIRONMENT IN BRITAIN AND IRELAND

Nigel H. Woodcock
University of Cambridge

First published in 1994 by UCL Press

Reprinted 1998, 2000

Taylor & Francis Ltd
11 New Fetter Lane
London EC4P 4EE

UCL Press is an imprint of the Taylor & Francis Group

The name of University College London (UCL) is a registered trade mark used by UCL Press with the consent of the owner.

ISBN: 1-85728-054-7

British Library Cataloguing-in-Publication Data

CIP data for this book is available from the British Library.

Typeset in Times Roman and Helvetica.
Printed and bound by
Bell & Bain Ltd, Glasgow.

PREFACE

This book is derived from a course that I teach to first year undergraduate students taking geology as part of the Natural Sciences degree in Cambridge. When I first prepared this course, I found that most available textbooks on environmental geology were published in the United States. Excellent though most of these books are, they focus on North American examples that seem remote to a student audience in the British Isles. When I came to compile more relevant data, I found them to be widely dispersed, some in publications not easily accessible to most students or to many of their teachers.

Geology and environment in Britain and Ireland is my attempt to outline the ways in which geology relates to human society, and to illustrate these themes with data and case studies from the British Isles. I have assumed some knowledge of basic geological principles, but no more than a student will acquire in an A-level Geology course or from a first-year subsidiary course at university level. The book can therefore complement or form the basis for taught courses on environmental geology at these levels, or can be used as supplementary reading by students interested in the relevance of geology to the environment. The book should also be useful to professional and amateur geologists in the British Isles and to other environmental scientists who need a compact geological perspective. The compilations of data and their sources provide a concise reference for scientists outside the British Isles needing statistics for comparative purposes.

The focus on the British Isles in this book provides a route into the wider field of environmental geology for readers familiar with this region. However, this focus is meant to obscure neither the general applicability of environmental geology, nor the global nature of some of its related problems. Accordingly, appropriate chapters contain both an introduction to relevant geological processes and a global perspective which broadens the view outwards from the British Isles examples. Suggestions for further reading on both regional and global issues are provided at the end of each chapter. Primary sources of all data are listed separately, so that these can be used for more detailed studies if necessary. Relevant data from the Irish Republic are given where possible, although such data are not always available or directly comparable with those from the UK.

Environmental geology is a broad subject, impinging by its very nature on many other disciplines. I have attempted to distil the subject into a short book that could be afforded by its intended readers. This has resulted in a high density of information on each page. However, I have used three devices to make both the text and figures clear and understandable. First, the purpose-drawn figures carry as much of the information as possible, some of which is not repeated in the text. Secondly, the figures always appear on the same page-spread as the relevant text, which is itself divided into page-fitted topics. Thirdly, I have provided a glossary of technical terms, including the words that appear in bold type in the main body of the book but which are not expicitly defined there.

To keep the book to its intended length, I have omitted some non-geological material which would have helped to put environmental geology in its full social context. Part I of the book delimits more precisely what has been included or omitted. The lists of further reading suggest ways of following up the non-geological themes.

The figures in this book have been specially designed and drafted. Sets of selected figures are available as coloured transparencies, as an aid to teachers or other lecturers. Details are available from the author at the Department of Earth Sciences, University of Cambridge, Cambridge CB2 3EQ, UK.

I have had much help and encouragement during the preparation of this book. The University of Cambridge and Clare College granted me a period of leave during which much of the book was drawn and written. The Department of Earth Sciences provided essential computing facilities, to whose complexities Ashley Meggitt and Eric Browne were willing guides. Ruth Banger and Dudley Simons gave efficient bibliographic and photographic help. Valuable comments at the outset of the project came from Philip Allen, Bob Allison, Derek Coussell and Dorrik Stow. Encouragement and advice during the book's preparation came particularly from Jeremy Leggett, Polly O'Hanlon, Douglas Palmer and Kevin Pickering. Chris Stillman and Tim Atkinson both read a complete draft of the book to a tight schedule, and suggested invaluable improvements. Roger Jones provided wise guidance throughout, and was especially tolerant of an author who wanted to design as well as write a book.

SOURCES OF ILLUSTRATIONS

All line drawings have been specially designed for this book. Those listed below have been adapted from a published source. The permission of the relevant individuals or organizations to use their material in this way is gratefully acknowledged.

Fig. 2.3b from Fig. 20 in Geological Museum (1988), © The Natural History Museum, London; Fig. 2.7a from Fig. 1 in Robinson & Williams (1984); Fig. 2.7b from section 2 on British Geological Survey (1984); Fig. 2.8 from Figs 17.1 and 17.4 in Boulton et al. (1977) by permission of Oxford University Press; Fig. 2.9 from Fig. 119 in Dunning et al. (1978), Crown copyright reserved; Figs 3.1b and 3.2 from Figs 5:8 and 4:6 in Dury (1973); Figs 3.3 and 3.6 from Figs 37, 32, and 36 in Watson & Sissons (1964); Fig. 3.4 from p. 19 in Penoyre & Penoyre (1978); Figs 4.2 and 4.21 from Figs 1.3 and 2.3 in Smith (1992); Fig. 4.10b from Fig. 3 in Cripps & Hird (1992); Figs 4.11, 4.15 and 4.16 from Figs 8.6, 4.3 and 4.13 in Perry (1981); Fig. 4.18b from Figs 1 and 2 in Cooper (1986); Fig. 9.7 from Fig. 6.5 in Taylor (1990); Fig. 10.2 from Fig. 13.8 in Guion & Fielding (1988); Fig. 10.3 from Fig. 8 in Fielding (1984); Fig. 10.6 from Fig. 3.1 in Brown & Skipsey (1986); Fig. 10.7 from Francis (1979); Fig. 11.14 from Fig. 12.1 in Parsley (1990); Fig. 11.15 from British Gas (1990); Fig. 11.18 after Gallois (1978); Fig. 11.19 from Macdonald (1990) adapted/modified with permission from the *Annual Review of Energy* volume 15, © 1990 by Annual Reviews Inc.; Fig. 14.2 from Fig. 15 in Porter (1978); Fig. 14.3 after Figs 5.2, 6.7, 6.11 and 7.10 in Waltham (1989); Fig. 14.4 from Fig. 4.3 in Waltham (1989); Fig. 14.5 from Fig. 3 in Wallwork (1956); Fig. 14.8 from Fig. 26b in Wallwork (1974); Fig. 15.3 from Fig. 1 in Gowring & Hardie (1968); Fig. 15.5 from Fig. 1 in Eyre (1973); Fig. 15.6a from Fig. 3.1 in Anderson & Trigg (1976); Fig. 15.6b from Fig. 3.69 in Lumsden (1992) by permission of Oxford University Press; Fig. 15.10 after British Geological Survey (1979) © Crown copyright; Fig. 15.12 from Fig. 7 in Steers (1981); Fig. 15.13 from Fig. 4.5 in Department of the Environment (1992); Fig. 15.14 from Figs 3 and 6 in Darby (1969); Fig. 16.3 from Figs 4 and 5 in Williams et al. (1984); Figs 16.6b and 16.7 from Figs 7 and 4 in Department of the Environment (1986) by permission of the Controller of HMSO; Fig. 16.8 from Figs 5.1 to 5.4 in UK Nirex (1987);

Those graphs and maps listed below are based on the cited data sources, the availability of which is gratefully acknowledged.

Fig. 3.1a from Watson & Sissons (1964); Fig. 3.7 from Johnston & Doornkamp (1982); Fig. 4.3 from Thompson (1982) as quoted in Smith (1992); Fig. 4.4 from Chicken (1975); Fig. 4.13 (stability data) from Lumsden (1992) and (erosion rates) from Goudie (1990); Fig. 4.20 from Department of the Environment (1992); Fig. 5.3 from White (1987); Fig. 5.4 from Skinner (1987); Fig. 6.5 from Avery (1990); Figs 6.6, 6.7 and 6.8 from Department of the Environment (1992), Cruickshank and Wilcock (1979) and Gillmor (1979); Fig. 6.9 from World Resources Institute (1992); Fig. 7.1 from L'vovich (1979); Figs 7.2, 7.3, 7.7, 7.9 and 7.10 from Department of the Environment (1992); Fig. 7.11 and 7.12 from World Resources Institute (1992); Figs 8.10 and 8.11 from British Geological Survey (1993); Fig. 8.12 from United Nations (1992); Figs 9.1, 9.2a and 10.9 from British Geological Survey (1993); Fig. 9.2c,d from Carr & Herz (1989); Figs 9.9, 9.10 and 9.11 from British Geological Survey (1992); Fig. 9.12 from Weisser et al. (1987); Fig. 10.5 from Carr & Herz (1989); Figs 10.10, 11.17, 11.21 and 11.22 (reserves) from British Petroleum (1993 and preceding years); Figs 11.2 and 11.3 from Cornford (1990); Fig. 11.16 from Brennand et al. (1990); Fig. 11.19, adapted/modified, with permission, from the Annual Review of Energy volume 15, © 1990 by Annual Reviews Inc.; Fig. 11.20 from Shell (1991); Fig. 11.22 (production and resources) from Masters et al. (1987, 1990); Fig. 12.1 from Darmstadter (1971) and British Petroleum (1993 and preceding years); Fig. 12.2 from World Resources Institute (1992); Fig. 12.3 oil, gas and coal reserves from British Petroleum (1993), coal resources from World Energy Conference (1986), oil and gas resources from Masters et al. (1990), unconventional resources from Masters et al. (1987); Fig. 12.4 from Bookout (1989); Fig. 12.7 from British Geological Survey (1992); Fig. 12.9 from Downing & Gray (1986); Fig. 12.10 from Energy Technology Support Unit (1985, 1987); Fig. 12.11 from Energy Technology Support Unit (1989); Fig. 12.12 from Flood (1986); Fig. 14.2a from Porter (1978); Fig. 14.7 from Wallwork (1974); Fig. 14.9 from Department of the Environment (1984, 1991) and Wallwork (1974); Fig. 14.10 from Department of the Environment (1992); Fig. 15.8 from Brookes et al. (1983); Fig. 15.14a from Waltham (1989); Fig. 15.16 from Brassington & Rushton (1987); Fig. 15.17 from World Resources Institute (1992); Figs 16.1 and 16.5 from Department of the Environment (1992); Figs 16.9 from World Resources Institute (1992); Fig. 17.2a, c, d from Department of the Environment (1992); Fig. 17.2(b) from Bailey et al. (1986); Fig. 17.3 from Nisbet

SOURCES OF ILLUSTRATIONS

(1991); Fig. 17.4 and 17.6 from Houghton et al. (1992); Fig. 17.5 and 17.7 from Houghton et al. (1990); Fig. 17.8a from Macdonald (1990); Fig. 17.9 from Tooley (1993) and Raynaud et al. (1988); Fig. 18.3 from the Graduate Geological Scientists' Surveys (Bottomley 1992 and preceding years);

All photographs are by the author. The cover illustration was provided by Earth Images, © DRA-Earth Images. A wide selection of images are available for reproduction or purchase as lithographic prints, photographs, data etc. Contact the Earth Images picture library, PO Box 43, Keynsham BS18 2TH. Tel/Fax 0275 839643.

I ENVIRONMENTAL GEOLOGY

A road bridge over the River Lune near Tebay, Cumbria. Its concrete pillars are founded on dipping massive sandstones of Silurian age – one of the more direct ways in which geology underlies human activity.

The human environment ranges beyond the present Earth's surface and its oceans and atmosphere. The environment extends downwards into the rocks of the subsurface and has a history that stretches backwards through time. These third and fourth dimensions are the concern of environmental geology – the subject which relates the science of the Earth to the human societies that the Earth supports.

The first chapter of this book previews the scope and purpose of environmental geology.

Chapter One

ENVIRONMENTAL GEOLOGY

1.1 Definition and content

Environmental geology relates the science of the Earth to the activities of human beings. Before the upsurge in environmental awareness of the past two decades, most of the concerns of environmental geology were grouped as **applied geology** – a contrast with "pure geology" which was taken to have no direct practical value. Most of applied geology could in turn be termed **economic geology** – the science of finding subsurface materials such as minerals and fossil fuels. However, environmental geology is more than just a fresh name for these older endeavours. First, it is more concerned than before with the ways in which geological processes influence the human environment, both through slow processes such as the influence on landscape, and through hazardous events such as earthquakes or landslides. Secondly, environmental geology focuses not just on the inputs to society but on its outputs: wastes and environmental degradation.

The concerns of environmental geology are most simply illustrated by describing the scope and organization of this book. It has a core in three parts, each dealing with a major interface between geology and human society (Fig. 1.1). Part II, *Geological influences on society,* covers the natural geological processes that influence the environment and its use. These processes would operate irrespective of human activities, although some human actions may trigger or accelerate natural processes such as floods, ground subsidence or earthquakes. Part III, *Earth resources for society,* describes the subsurface resources that geologists are called upon to find. These include land, water and industrial minerals as well as coal, petroleum and metallic minerals. The exploitation of these resources has its own environmental impacts, discussed in Part IV, *Human impacts on the Earth.* This part also covers the deliberate impacts of engineering and agriculture, and the unplanned pollution of land, water and air by the wastes from society.

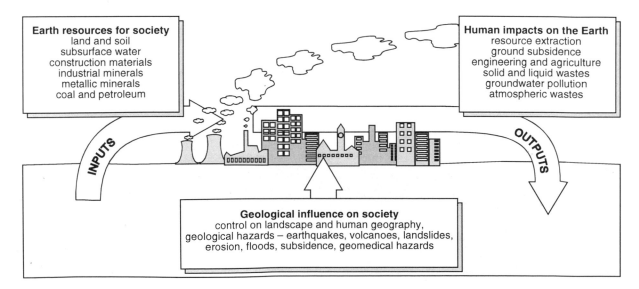

Figure 1.1   The components of environmental geology used as a basis for this book.

The three central parts of the book are set in a broader context by the introductory Part I, *Environmental geology,* and the concluding Part V, *Geology and society.* These chapters review the place of geology within the environmental sciences, and within society in general. Most of this discussion is deferred until Part V, which can then draw on the specific examples in the main body of the book. However, some important themes must be outlined here at the outset. Particularly important are the constraints on the content of this book – the thinking behind what material to include and, as important, what to omit.

This book is not about **environmental science** in general, but about its "geological" or "Earth science" component. A view of the relationships between these disciplines (Fig. 1.2) shows environmental science as the study of the total human environment on the present Earth. This environment is simplified as a series of concentric spheres: the mainly rocky **lithosphere**, the watery **hydrosphere**, the organic **biosphere**, the mainly gaseous **atmosphere** and its surrounding **magnetosphere. Earth science** is dominated by the study of the lithosphere and the inner parts of the present Earth. It should not be confused with **Earth systems science**, which emphasizes the interactions of all environmental components, and is broadly synonymous with environmental science.

In this view, **geology** is the Earth systems science of the past Earth. It has the crucial extra dimension of geological time, and the aim of reconstructing the history of the Earth and its mechanisms. So geologists are interested in active processes at and above the Earth's surface as well as below it, to use as a guide to similar processes in the past. Geological history is also proving an important guide to predicting future changes in the global environment, such as fluctuating climate or rising sea levels.

Guided by these definitions, this book will concentrate on the environmental lessons to be learned from Earth history, and particularly on the subsurface environment of the present Earth. This geological perspective is necessary to expand the prevalent popular view of "the environment", which neither reaches below the Earth's surface nor extends back through time. Only in one chapter might the book's content seem to stray outside its geological brief – that on atmospheric wastes (Ch. 17). The reasons will become clear. Most of these wastes are produced by burning the fossil fuels that geologists have found, and their threatened impacts on the global climate can be matched by examples from the geological past.

The geological emphasis of this book is not meant to imply that environmental geology can be fully understood in isolation from other environmental sciences or from the social context in which it is learned and practiced. Indeed, Part V of this book will support the

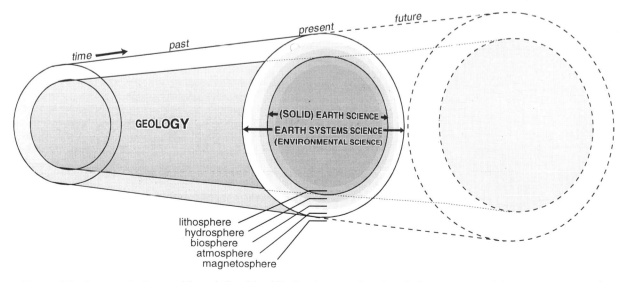

Figure 1.2 A conceptual view of the relationship of Earth science and geology to the environmental sciences in general.

3

case for a more integrated, holistic approach to the study of the Earth (Fig. 1.3). More important still, it will argue that the environmental sciences must interact closely with the societies in which they operate, accepting that science is influenced by, and able to influence in turn, the political, social and economic climate. The idea that science is totally objective is particularly suspect in this area of human knowledge. However, Parts II, III and IV of this book make a deliberate attempt to describe the factual elements of environmental geology in an impartial and dispassionate way. Only in Part V will some of the ethical and sociological threads that join these elements be drawn together.

This book limits itself not only to a geological content, but to illustrating that content mainly with examples from the British Isles. This is intended to be an enabling strategy for the readers, rather than one that narrows their perspective. Many environmental impacts are primarily local or regional in scale, and are best understood in relation to examples from a familiar geographical context. Global reverberations of these impacts can then be better appreciated. Most chapters in this book therefore begin with an explanation of general principles, applicable anywhere in the world. Only then are these principles illustrated using case studies from the British Isles. Where appropriate, chapters end with a global perspective that highlights where environmental concerns differ from those in western Europe. The list of further reading provides further examples of both European and North American perspectives on environmental geology.

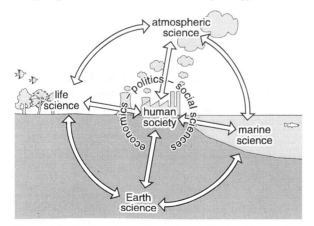

Figure 1.3 Earth science shown in relation to other environmental sciences and to their influence by society.

Further reading

Bowler, P. J. 1992. *The Fontana history of the environmental sciences.* London: HarperCollins. [A useful review, with geology properly featured, but stronger on nineteenth century and earlier history than on recent developments.]

Blunden J. & A. Reddish 1991. *Energy, resources and environment.* London: Hodder & Stoughton/Open University. [The most relevant of a series of books based on Open University courses, and emphasizing case studies from the British Isles.]

Chiras, D. D. 1991. *Environmental science: action for a sustainable future.* [Good example of the many stimulating introductions to environmental science, mostly North American, and mostly with a biological rather than geological emphasis.]

Coates, D. R. 1981. *Environmental geology.* New York: Wiley. [Probably the most thoughtful and rewarding example of North American books on geology and the environment.]

Department of the Environment 1992. *The UK environment.* London: HMSO. [A useful though selective digest of UK government statistics on the environment, with an introduction to relevant processes.]

Keller, E. A. 1992. *Environmental geology,* 6th edn. New York: Macmillan. [One of the first North American textbooks on environmental geology. Well illustrated, with a short but adequate introduction to general geological principles.]

Lumsden, G. I. (ed.) 1992. *Geology and the environment in western Europe.* Oxford: Oxford University Press. [A valuable compilation of European case studies with some introduction to the geological principles behind them.]

Montgomery, C. W. 1992. *Environmental geology,* 3rd edn. Dubuque, Iowa: Brown. [A well-illustrated introduction to the full scope of geology, with later environmental sections based on North American examples.]

Stow, D. A. V. & D. J. C. Laming 1991 *Geosciences in development.* Rotterdam: Balkema. [Collected case studies that provide a resource-led view of environmental geology in the developing world.]

Watson, J. 1983. *Geology and man: an introduction to applied Earth science.* London: Allen & Unwin. [A clearly written introductory level textbook emphasizing examples from the British Isles.]

White, I. D., D. N. Mottershead & S. J. Harrison 1992, *Environmental systems: an introductory text,* 2nd edn. London: Chapman & Hall. [An excellent guide to the systems science approach to the environment, emphasizing examples from the British Isles.]

II GEOLOGICAL INFLUENCES ON SOCIETY

WARNING
BEWARE OF CLIFF FALLS
YOU ARE ADVISED
TO KEEP CLEAR
OF THE CLIFFS

THE OWNERS ACCEPT
NO LIABILITY.

Cliffs of Upper Cretaceous Chalk west of Lulworth, Dorset, retreating at up to 40 cm per year by undercutting and rockfall. Rapid coastal erosion is a geological hazard along many parts of the south and east coasts of England.

The environment of the British Isles is typically seen in terms of its surface components: climate, water, air vegetation, wildlife, and the human modifications to the built environment. However, subsurface geology also influences the natural environment and the ways in which humans use, alter and exploit it. The first part of this book reviews these geological influences on society.

Chapter 2 describes the geological controls on the natural landscape of the British Isles, in particular the divide between the uplands of the north and west and the lowlands further southeast.

Chapter 3 considers the human settlement and activity in this natural landscape. Patterns of population, agriculture, industry and even traditional architecture are all geologically influenced. The distribution of natural resources is considered in more detail in Part III.

Most geological influences on society are benign and passive. By contrast, geological hazards are active, if in some cases remote, threats to society. Chapter 4 assesses the risk from landslides, floods and even earthquakes in the British Isles.

Chapter Two

GEOLOGY AND LANDSCAPE

2.1 Geological history of the British Isles

The varied rocks and landscape of the British Isles result from their position at one of the Earth's geological crossroads (Fig. 2.1). Seven large fragments of continental crust influenced the Phanerozoic geological history of this area, even before the Mesozoic formation of the Atlantic Ocean. This section briefly reviews this history in terms of a sequence of crustal shortening and extension events (Fig. 2.2). This event framework largely determines the succession and distribution of rock units in the British Isles (Ch. 2.2)

Until Silurian time, the north and south of the British Isles were geographically separate. Scotland and northwest Ireland were part of **Laurentia** in low southern latitudes (Fig. 2.2c). These areas had a complex early history dating back at least 2 000 million years before the Cambrian period. The **Grampian orogeny** was merely the last of a series of Precambrian events that deformed and metamorphosed the Laurentian crust (Fig. 2.2a). The **Athollian orogeny** further heated and uplifted the Laurentian continental margin in Ordovician time.

England, Wales and southeast Ireland were part of the major **Gondwana** continent in late Precambrian time, when they were affected by the **Cadomian orogeny**. By the Cambrian, this part of Gondwana lay in high southern latitudes, 40° away from northern Britain across the intervening **Iapetus Ocean**. In early Ordovician time southern Britain began to rift away from Gondwana as part of the **Eastern Avalonia** crustal fragment, probably accompanied by **Western Avalonia**, **Iberia**, **Armorica**, and possibly **Baltica**. Eastern Avalonia drifted northwards, narrowing the Iapetus Ocean which separated it from the Laurentian continent.

This northward drift was accompanied by spasmodic crustal extension within Eastern Avalonia. However, in Silurian time it began to impinge on Laurentia, and crustal shortening resumed. This collision culminated in Devonian time with the **Caledonian** or **Acadian orogeny**, deforming and weakly metamorphosing all the British Isles except northwest Scotland and southern England.

By late Devonian time Eastern and Western Avalonia and Baltica were welded to Laurentia, but still separated from Gondwana and related fragments to the south by the **Rheic Ocean.** After Carboniferous extension, the closure of this ocean was accompanied by the **Variscan orogeny**, affecting the southern fringe of Ireland, Wales and England (Fig. 2.2a).

By now the British Isles had drifted across the Equator (Fig. 2.2c) and it continued to move northwards as part of the global supercontinent **Pangaea**. This continent began to break up in Permo-Triassic time, initiating new oceans such as the Atlantic. The associated crustal extension dominated the British area through Triassic to Palaeogene time, particularly in the North Sea and on the Atlantic margin (Fig. 2.2a).

The weak **Alpine shortening** affected southern England in the Palaeogene, a remote effect of continental collision in southern Europe. Subsequent crustal movements have been dominantly vertical (Ch. 2.2).

Figure 2.1 The main continental fragments relevant to the geological history of the British Isles.

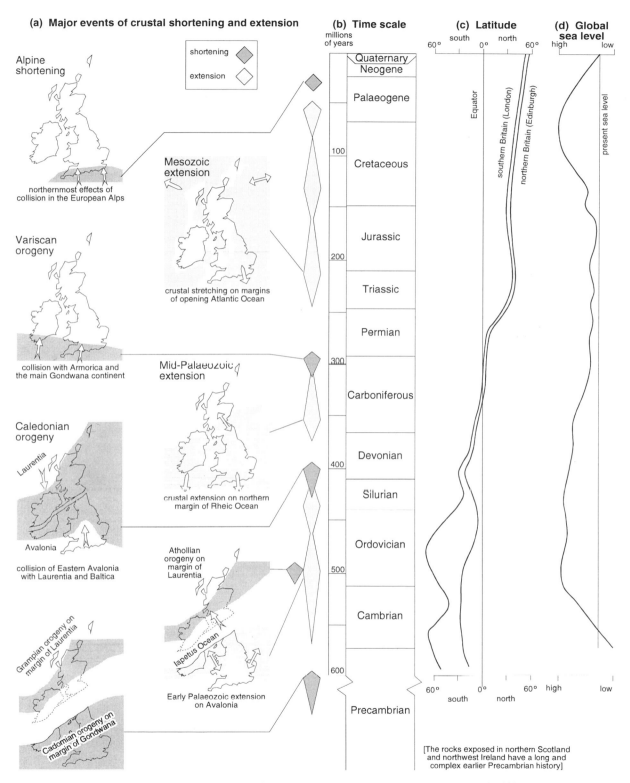

Figure 2.2 Tectonic events, latitude and sea level changes in British Isles geological history.

2.2 The geological map of the British Isles

Landscape variations are less directly influenced by geological history than by the present pattern of rock units at the Earth's surface. The geological map of the surface of the British Isles (Fig. 2.3b) is partly controlled by the spatial pattern of geological events (Fig. 2.2) and of the rock units that they produced (Fig. 2.3a). So rocks of the Variscan orogenic belt are necessarily restricted to the southern fringe of Britain and Ireland. However, just as important is the more recent pattern of uplift and subsidence (Fig 2.4).

The rock sequence in the British Isles (Fig. 2.3a) mirrors clearly the events that formed it (Fig. 2.2a). Episodes of crustal shortening are marked by widespread **unconformities**, gaps in the geological sequence recording the uplift and erosion associated with **orogeny** or mountain building. Periods of crustal extension are recorded by accumulations of sedimentary or volcanic rocks on subsiding crust. **Igneous activity** can accompany both extension or shortening of the crust. Major **metamorphic events** coincide only with strong shortening and so affect the rocks below each widespread unconformity.

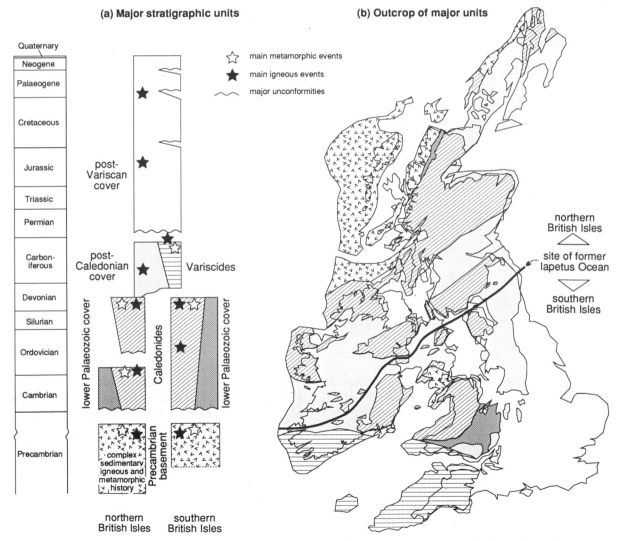

Figure 2.3 Major stratigraphic units and their outcrop pattern in the British Isles. Part b is reproduced with permission from *Britain's offshore oil and gas,* Geological Museum, 1988; © The Natural History Museum, London.

8

The most significant unconformity in the British Isles is that in Devonian time resulting from the **Caledonian orogeny**. Only then did the geological histories of northern and southern Britain converge as the intervening Iapetus Ocean finally closed. The presence of a major late Precambrian unconformity in both regions is a coincidence.

The southern British Isles are underlain by **Precambrian basement** no older than 700 Ma. These rocks outcrop extensively only in northwest Wales and southeast Ireland (Fig. 2.3b). In south and east Wales and the English Midlands they protrude as small inliers through the later cover. The Precambrian basement was metamorphosed during the Cadomian orogeny, then eroded and overlain by a **lower Palaeozoic cover**. In southeast Wales and central England this cover remained largely unaffected by the Caledonian orogeny. Elsewhere it was strongly deformed and weakly metamorphosed to produce the southern **Caledonides**. These outcrop in southeast Ireland, Wales and northwest England, but also underlie much of eastern England.

The Precambrian rocks of the northern British Isles record a longer and more complex history than further south, dating back nearly to 3000 Ma. Geological histories vary markedly from area to area across major faults. No attempt is made to subdivide this history on Figure 2.3, which shows only the Grampian orogeny, the latest of the Precambrian events to affect the northern British Isles. **Precambrian basement**, including some unmetamorphosed sedimentary rocks, outcrops extensively in northwest Scotland. Overlying it is a remnant of **lower Palaeozoic cover**, also only weakly affected by later metamorphism. However, over most of Scotland and northwest Ireland, Precambrian and lower Palaeozoic rocks were subjected to the Athollian orogeny in Ordovician time and then by the Caledonian metamorphism and deformation. The **Caledonides** of southern Scotland and their continuation into northeast Ireland suffered only low metamorphic temperatures, but further northwest higher grades were reached.

The **post-Caledonian cover** blanketed much of the British Isles, and is known to underlie the southern North Sea. Over most of this area it is unmetamorphosed and only weakly deformed, but in the south it was strongly affected by the Variscan orogeny. The

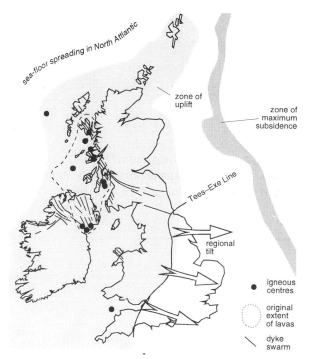

Figure 2.4 Palaeogene uplift and subsidence in relation to igneous activity.

resulting **Variscides** outcrop in southern Ireland, south Wales and southwest England, but continue eastwards beneath southern England (Fig. 2.3b).

The **post-Variscan cover** was deposited particularly over southeast Britain and the North Sea, and in a series of mostly offshore basins in the west of the British Isles. Many local unconformities interrupt the sequence (Fig. 2.3a) but no metamorphic or deformation events of Variscan or Caledonian intensity. However, weaker events accompanying the Palaeogene opening of the north Atlantic Ocean had a strong influence on the geological map of Britain (Fig. 2.4). Rifting was accompanied by igneous activity on the adjacent continental margins, probably above an anomalous **mantle hot spot**. This activity heated the crust, lowering its density and enhancing the uplift of the margin. While north and west Britain were rising, the North Sea crust was continuing to subside in the aftermath of its earlier extension. The result was a west-to-east tilt of the British Isles, and subsequent erosion down to older rocks north and west of the **Tees–Exe Line** (Fig. 2.4).

2.3 Landscape and solid geology

Landscape in the British Isles is primarily influenced by the underlying "solid", that is pre-Quaternary, geology. Other important factors are variations in climate and particularly in processes during and since the Quaternary glaciations (Ch. 2.4).

Rocks have widely differing resistance to the landscape-forming processes of weathering and erosion. This resistance arises from the nature of their constituent particles and from the ways in which the particles are bound together. Every stratigraphic sequence has its own detailed character but, in general, older sedimentary or metasedimentary rocks are more resistant than younger rocks (Fig. 2.5). Old rocks have had

the pores between particles closed up by **compaction** and filled by **cementation**. They are more likely to have recrystallized during **metamorphism**. Most igneous rocks, of whatever age, are formed with a tight crystalline structure that resists erosion and weathering (Fig. 2.5).

When the north and west of the British Isles were uplifted in Palaeogene time (Ch. 2.2) its veneer of soft Mesozoic rocks was rapidly eroded. Erosion rates decreased as Carboniferous and older rocks reached the land surface, and uplands and mountains developed. The Tees–Exe Line bounds the northwestern region containing nearly all outcrop of pre-Permian rocks (Fig. 2.3) together with most of the upland areas of the British Isles (Fig. 2.6).

Lowland scenery reflects not only the weakness of the whole rock sequence but also the marked contrasts in strength between one unit and another. **Mudstones** or **shales** generally form the weakest rock units in the British Mesozoic and Cenozoic. **Sandstones** and **limestones** are stronger, unless they have remained uncemented. Alternations of these rock types produce **scarp-and-vale topography**. The anticlinal structure

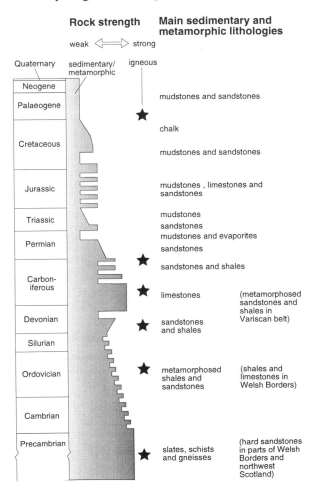

Figure 2.5 Generalized plot of rock strength against stratigraphic age for typical rock types in the British Isles.

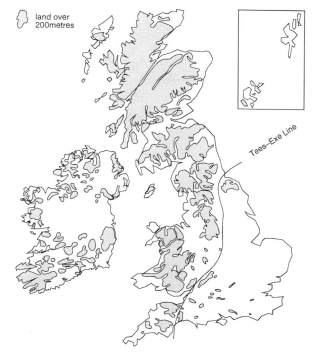

Figure 2.6 Uplands and lowlands in the British Isles.

of the southeast of England provides an example (Fig. 2.7a). The resistant Cretaceous Chalk forms the North and South Downs, in continuity with low Chalk ridges and hills across the English lowlands. The clay-dominated sequence below the Chalk forms lower ground, interrupted by weak ridges above interbedded sandstone units. Sandstones of the Ashdown Beds underlie the High Weald in the core on the anticline. Elsewhere, the Jurassic limestones and shales form particularly persistent scarp-and-vale topography from the Cotswolds northeastwards through eastern England to the North York Moors, where sandstones add to the variety and topography.

Upland scenery depends strongly on whether the underlying rocks have been deformed or metamorphosed, or form part of the sedimentary cover that escaped these orogenic events. The weakly deformed cover rocks can form scarp-and-vale topography similar to that on Mesozoic rocks. Examples occur on the Silurian and Devonian rocks of the Welsh Borderland and on the Upper Carboniferous sandstones and shales of the coalfields.

Deformation and metamorphism subdue the stength contrasts between sedimentary units. Landscapes developed on the rocks of the Variscan, Caledonian and earlier orogens therefore show less conspicuous lithological control than in the lowlands. In mid-Wales, for example (Fig. 2.7b), higher ground can develop just as commonly on mudstones as on sandstones. The susceptibility to weathering is influenced as much by the strength of tectonic cleavage as by lithology. This topographic disregard for rock type is even more marked in the high-grade metamorphic areas in the north and west of Scotland and Ireland.

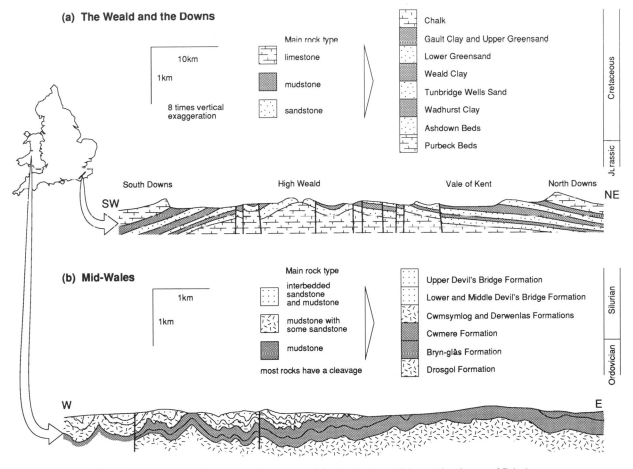

Figure 2.7 Geological cross section across (a) a lowland and (b) an upland area of Britain.

2.4 Landscape and Quaternary geology

About 2.4 million years ago, climatic deterioration initiated the alternations of glacial and interglacial conditions that have continued to the present day. Glacial erosion and deposition strongly embellished a landscape already established by Paleocene uplift (Ch. 2.3). In the British Isles, glacial effects are confined to the past 700 000 years of Quaternary time. Particularly important were the extensive Anglian glaciation – 450 000 years ago – and the more recent Devensian event – 18 000–17 000 years ago (Fig. 2.8).

Ice accumulated most rapidly in the already established uplands of Scotland, western Ireland, the English Lake District and north Wales. Ice sheets flowing out from these uplands merged with each other and with the front of the Scandinavian ice sheet, spreading over the Atlantic margin and into the British lowlands (Fig. 2.8). Erosion was favoured below fast-moving ice, and deposition below slow-moving ice, especially near its melting point.

The uplands have suffered the most intense glacial erosion (Fig. 2.8a), with major **troughs**, **cirques** and **scoured rock surfaces**. This erosion occurred when

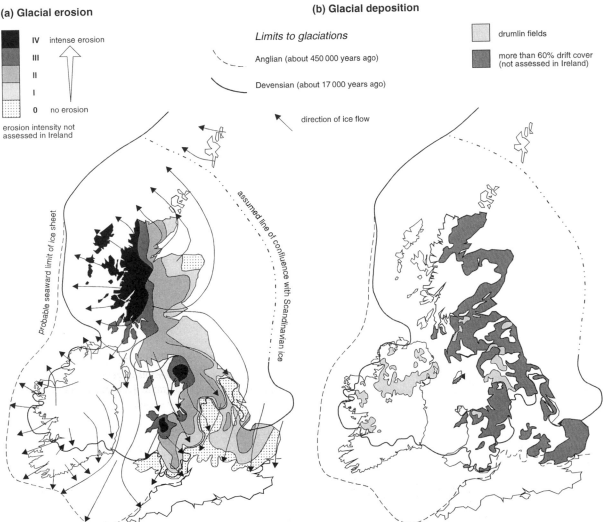

Figure 2.8 Distribution of glacial erosion and deposition in the British Isles.

the uplands hosted small, fast-moving valley glaciers, rather than the thick but slow-moving centres of major ice-sheets. These large ice-sheets flowed fastest near their margins in the lowlands. They scoured out deep troughs in areas such as the Dee and Mersey estuaries and the Wash, now filled with later sediment (Fig. 2.8a).

Glacial deposits are most extensive away from upland areas, where the ice-sheets were warming, melting and dropping their eroded debris as **till** or as sediment reworked by meltwater (Fig. 2.8). The deposits from the Devensian glaciation form identifiable landscape features such as **moraines** of till, or **kames** and **eskers** formed of outwash sediment. **Drumlins** are swarms of elongate hills made of moulded till. They occur in extensive areas of Britain and Ireland (Fig. 2.8b), where large volumes of glacial debris were melting out and ice velocities were high enough to mould this debris into shape.

The deposits from older glaciations typically form featureless sheets, their original landforms now degraded by **periglacial** activity. These processes, such as seasonal freezing and thawing, naturally affected the British Isles even south of the ice margin, and continue to the present day in the high uplands.

The weight of thick Quaternary ice depressed the crust, particularly in northern Britain. Since the most recent permanent ice sheet melted about 10 500 years ago, the crust has been rebounding (Fig. 2.9). The resulting pattern of uplift and subsidence has a continuing effect on landscape, particularly at the coast (Ch. 4.5).

Further reading

Anderton, R., P. H. Bridges, M. R. Leeder, B. W. Sellwood 1979. *Dynamic stratigraphy of the British Isles: a study in crustal evolution*. London: Allen & Unwin. [A stimulating account of geological history, integrated with plate tectonic models.]

Bowen, D. Q. 1978. *Quaternary geology*. Oxford: Pergamon. [A readable introduction to Quaternary processes and their effects in Britain.]

Craig, G. Y. 1991. *Geology of Scotland*, 3rd edn. London: Geological Society. [This and the next two books are good detailed guides to the geology of the British Isles.]

Duff, P. McL. D. & A. J. Smith 1992. *Geology of England and Wales*. London: Geological Society.

Holland, C. H. (ed.) 1981. *A geology of Ireland*. Edinburgh: Scottish Academic Press.

Dunning, F. W., I. F. Mercer, M. P. Owen, R. H. Roberts, J. L. M. Lambert 1978. *Britain before Man*. London: HMSO. [A simple, reliable geological history of Britain.]

Goudie, A. 1990. *The landforms of England and Wales*. Oxford: Basil Blackwell. [An authoritative analysis of landscape development, particularly informative on important geomorphological processes.]

Goudie, A. & R. Gardner 1992. *Discovering landscape in England and Wales*. London: Chapman & Hall. [A well-illustrated introduction to landscape and its controls.]

Rice, R. J. 1988. *Fundamentals of geomorphology*, 2nd edn. Harlow, Middx: Longman. [A clear introduction to the principles of landscape evolution.]

Shotton, F. W. (ed.) 1977. *British Quaternary studies: recent advances*. Oxford: Clarendon. [A useful collection of papers giving an overview of most aspects of the glacial and postglacial geology of Britain.]

Whittow, J. B. 1974. *Geology and scenery in Ireland*. London: Penguin. [This and the next book give a systematic account of the scenery of the British Isles.]

Whittow, J. B. 1992. *Geology and scenery in Britain*. London: Chapman & Hall.

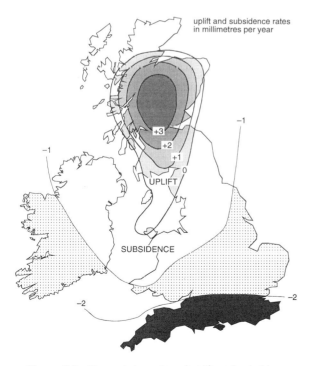

uplift and subsidence rates in millimetres per year

Figure 2.9 Present-day rates of uplift and subsidence.

Chapter Three
GEOLOGY AND HUMAN GEOGRAPHY

3.1 Rural geography

If the large-scale features of the landscape in the British Isles are derived from its varied geological history (Ch. 2), so too is much of its detail. Most of these superficial features are not natural but have been fashioned by human activity, itself influenced or constrained by geological factors.

Rural land accounts for more than 90% of the area of the British Isles (Ch. 6). About two-thirds of this land is improved farmland or managed forest and woodland. The rest is unimproved moorland or heath. The distribution of these two types of land-use is strikingly controlled by the upland/lowland divide along the **Tees–Exe Line** (Fig. 3.1a). Almost all the moorland occurs to the northeast of this line, on Carboniferous and older rocks. The main exception is the Jurassic area of the North York Moors. These uplands have poorer soils and harsher climates, making them unsuitable for growing crops or grass. The sparse human development in these areas has in turn given their scenery a special cultural value, so that most of Britain's national parks have been established in the uplands (Fig. 3.1a).

Improved farmland dominates the English lowlands, and has encroached over the lower ground in the upland zone. Therefore, farmed land dominates on the Permo-Triassic rocks of the Cheshire Plain, and on the Carboniferous of the Central Irish Plain and the Scottish Midland Valley. However, the use of this farm-

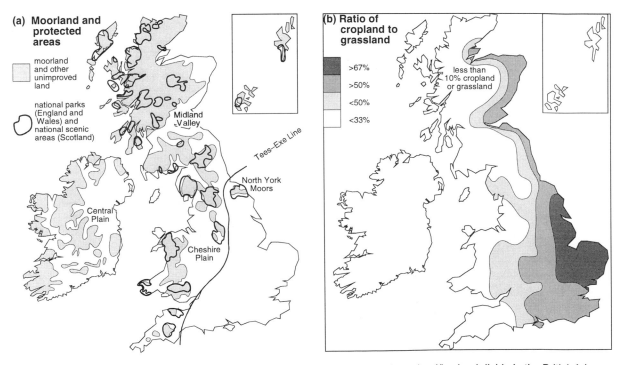

Figure 3.1 The pattern of national parks and of agriculture in relation to the upland/lowland divide in the British Isles.

land differs in the west and east of the British Isles (Fig. 3.1b). Grassland for livestock farming dominates the upland zone, whereas arable cropland is more common in the lowlands. This distribution results mainly from the predominantly wet westerly air flows dropping most of their rain over the western uplands. Grass grows best in the wetter west, with its mild winters. Grain crops do better in the lower rainfall and warmer summers of the lowlands.

The geological map of the British Isles has influenced the pattern of rural development on a local as well as a regional scale. For example, the clay vales of the English lowlands were not generally occupied by the original settlers (Fig. 3.2). These valleys were more prone to flooding, and their soils were more difficult to work. They were left as natural woodland or pasture. The foot of limestone or sandstone scarps provided the best sites for villages, with drier ground, more workable soil and often a supply of springwater. The scarp-top land afforded rough grazing for livestock, and was often a regional route of communication.

Geological influences on rural settlement abound throughout the British Isles. For instance, the Carboniferous scarp-and-vale topographies of the Pennines and the Scottish Midland Valley exert a control similar to that in the Mesozoic lowlands. Individual drumlins in Ireland provide better drained land for farms and fields. Even the detail of the rural landscape may

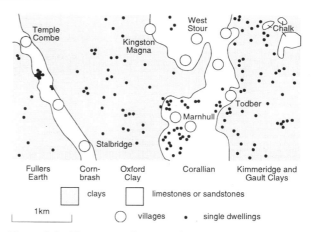

Figure 3.2 The pattern of present-day settlement in relation to underlying solid geology, west of Shaftesbury, Dorset.

reflect its geology through traditional buildings and walls (Fig. 3.3). Fields are usually edged with dry stone walls wherever suitable rock is available. However, stone is rarely transported far for so mundane a purpose, so that transitions from walls to fences or hedges often faithfully follow geological boundaries. Moreover, hedges grow best on moist nutrient-rich clay soils. Chalk uplands or heaths on soft sandstones provided neither the rich soils suitable for traditional agriculture nor the resistant stone necessary for walls. Such areas have been enclosed relatively recently by wooden or wire fences (Fig. 3.3).

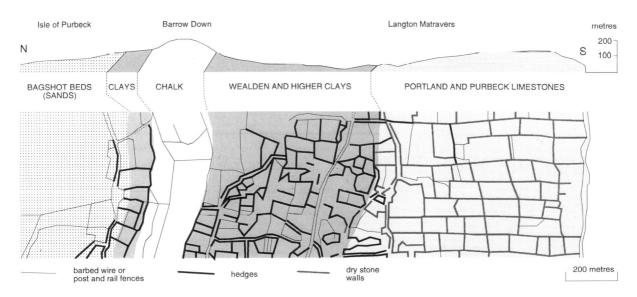

Figure 3.3 Types of field boundary in relation to geology in a north–south traverse across south Dorset.

(a) Main walling materials

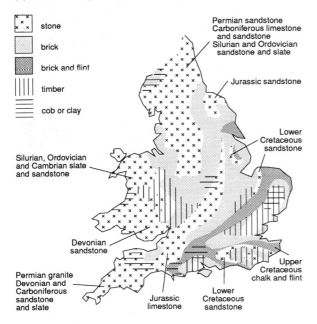

stone
brick
brick and flint
timber
cob or clay

Permian sandstone
Carboniferous limestone
and sandstone
Silurian and Ordovician
sandstone and slate

Jurassic sandstone

Lower
Cretaceous
sandstone

Silurian, Ordovician
and Cambrian slate
and sandstone

Devonian
sandstone

Upper
Cretaceous
chalk and flint

Permian granite
Devonian and
Carboniferous
sandstone
and slate

Jurassic
limestone

Lower
Cretaceous
sandstone

(b) Main roofing materials

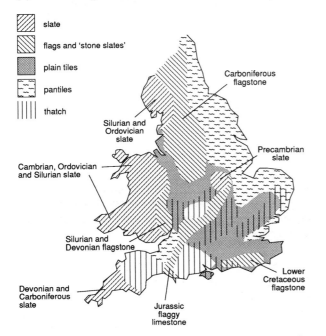

slate
flags and 'stone slates'
plain tiles
pantiles
thatch

Carboniferous
flagstone

Silurian and
Ordovician
slate

Precambrian
slate

Cambrian, Ordovician
and Silurian slate

Silurian and
Devonian flagstone

Lower
Cretaceous
flagstone

Devonian and
Carboniferous
slate

Jurassic
flaggy
limestone

Figure 3.4 Distribution of the dominant walling and roofing materials in rural vernacular buildings in England and Wales.

3.2 Vernacular architecture

Dry stone walls are the most humble type of **vernacular architecture**, a style of building that uses local materials with traditional designs and techniques. Vernacular buildings vary more between regions than through time, reflecting differences in culture, climate and, above all, construction materials. Because rocks provide many of these materials, vernacular architecture is perhaps the most conspicuous geological imprint on the human geography of the British Isles. This influence is shown by distribution maps of the traditional walling and roofing materials in England and Wales (Fig. 3.4). On this scale, and on any more detailed scale, the correspondence with geological maps is striking.

Stone is the main walling material over much of the uplands (Fig. 3.4a). The Pennines have Carboniferous sandstones or limestones, with Permian sandstones or limestones on their flanks. Lower Palaeozoic rocks dominate in the Lake District and Wales, mostly slaty mudstones. Devon and Cornwall have similar low-grade metamorphic rocks, supplemented by the Variscan granites.

The English lowlands have less abundant walling stone than the uplands. However, Jurassic limestone is widely used along its outcrop from Dorset to Lincolnshire, Jurassic sandstones in north Yorkshire, and Lower Cretaceous sandstones in eastern England. The Upper Cretaceous chalk is only sporadically hard enough to build with, but contains flint nodules which are durable, if difficult, to use.

Outside the stone belts, most lowland areas have clays suitable for baking as **bricks**, for making into unbaked **clay lumps**, or for compacting with straw to mould into **cob** walls. **Timber** was used for walls as well as roof supports where available, forming a frame infilled with brick or **wattle and daub**, wooden laths coated with clay, dung and horsehair.

Roofing materials in the upland zone are of two distinct types. Local areas of the Caledonian belt in the Lake District and Wales and of the Variscan belt in Cornwall furnish good quality **slates**, which split into thin sheets along their secondary tectonic cleavage. In the Pennines, some Carboniferous sandstones split along their primary sedimentary lamination into **flags**, which provide durable, though heavy, roofing mate-

Figure 3.5 Walling in (a) Lake District (Ordovician slate and Permian sandstone); (b) north Norfolk (Cretaceous red chalk, flint and brick), (c) north Yorkshire (Carboniferous sandstone) and (d) Devon (Devonian slate).

rial. Silurian and Devonian flags are used in the Welsh Borderland. In the lowlands, some flaggy Jurassic limestones and Cretaceous sandstones provide "**stone slates**", but **tiles** made from baked clay are more common. **Pantiles** are corrugated tiles to a design imported from the Low Countries and therefore most evident in eastern England. **Thatch** of reed, straw or heather was used where other roofing materials were unavailable or expensive.

The type of construction material influences not only the general appearance of a building, but also its architectural details. Examples of walling styles reveal this subtle geological influence (Fig. 3.5). A major design problem is to achieve strength despite the irregular shapes of broken or water-rounded stone. Corners and window or door openings are particularly vulnerable areas. Naturally regular stone, soft cut

stone (Fig. 3.5a) or bricks (Fig. 3.5b) were often brought in for these positions. Stone walls are usually made of two separate faces, filled with small rubble, but long **throughstones** are built in at intervals to bind the two faces together (Fig. 3.5c). The interlocking of rounded stones is improved by **galletting**, small stones pushed into the irregular joints (Fig. 3.5b). Irregular slaty rock is laid upright in zig-zag courses in some dry stone walls, so that the stones interlock better as the wall settles (Fig. 3.5d).

The detailed pattern of vernacular building materials in Scotland and Ireland is less well documented. Stone naturally predominates in surviving buildings, although **turf** used to be an important material for both walls and roofs. The traditional lime-washing of buildings tends to obscure a variety of walling materials comparable at least with that of Wales.

3.3 Urban and industrial geography

Urban areas occupy only about 8% of the land area of the British Isles. However, because they contain about 80% of the population and most industrial sites, the location of urban areas is a central component of the region's human geography and environment.

The locations of most towns and cities are influenced by their local physical geography, itself related to the underlying geology. For instance, the largest national cities of London, Glasgow, Cardiff, Belfast and Dublin all originated as ports at the focus of natural inland transport routes (Fig. 3.6b). Most other large coastal towns are also ports, but the inland conurbations of northern and central England are more difficult to understand without a knowledge of the underlying geology. For all these industrial and population centres, and those of central Scotland, northeast England and south Wales, are sited precisely on the British coalfields.

Before the Industrial Revolution (that is before about 1750) London was the only large town in the British Isles by modern standards. Two-thirds of the region's labour force were employed in agriculture, and much of the industrial activity was dispersed in rural areas. Ironworking, for instance, exploited local ores in the Jurassic of the Weald and in the Carboniferous coalfields, but its viability was limited by the shortage of wood to make charcoal for the smelting process. Coal had proved too sulphurous for smelting, making the iron too brittle. The key technological breakthrough, by the Darby family in the West Midlands of England, was first to bake the coal and produce sulphur-free **coke**. The coke–iron smelting process focused the ironmaking industry on the exposed Carboniferous coal sequences with their geologically associated iron-stone horizons. By the middle of the 19th century, all the exposed British coalfields were being exploited (Fig. 3.6a) and over half the population of England had migrated to fast-growing urban areas.

The other major factor that concentrated industry and its workforce on the coalfields was the development of the steam engine. The inefficient designs of the 18th century were improved by Smeaton and Watt to lead the mechanization of industry during the 19th century, a growth fuelled mostly by coal.

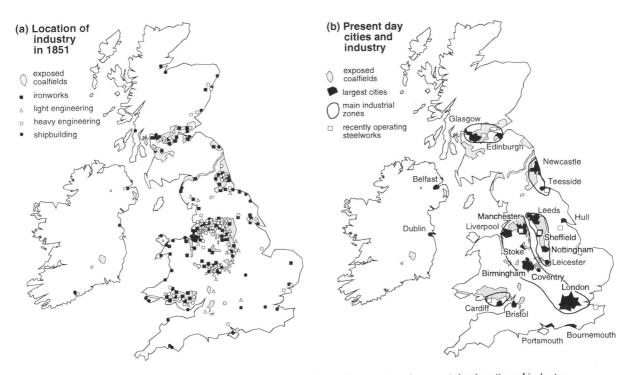

Figure 3.6 Carboniferous coalfields compared with the historical and present day location of industry.

A survey of industry at the mid-point of the 19th century shows the striking concentration of iron foundries and engineering works on the exposed coalfields (Fig. 3.6a). Shipbuilding was the only heavy industry to flourish beyond the coal areas, the need for deepwater sites outweighing the proximity of coal and, later, iron and steel supplies. The optimum shipyard sites were estuaries where the coalfields come close to the coast: the Clyde, Tyne and Mersey.

The other main industry that came to depend on coal-fired steam engines was the manufacture of cotton and woollen textiles. The siting of early textile mills was determined by the availability of water for power and for washing wool, and influenced by the availability of the fuller's earth – montmorillonite clay (Ch. 9.1) – used for cleansing. Increasing mechanization in the 19th century favoured mills close to coal supplies, and the industry gravitated to the Yorkshire and Lancashire coalfields. Here the soft water running off the Upper Carboniferous sandstones and shales of the Pennines was traditionally favoured for washing textiles, and the humid climate prevented threads breaking during manufacture.

In such ways, the geological map of Britain provided a template for the urban concentrations of the industrial revolution. This pattern has been blurred during the 20th century but is still a dominant geographical feature (Fig. 3.6b). Its persistence results from **industrial inertia**, where modernized or new industries tend to use established sites and workforces, even though the optimum location, based on transport or materials factors, may be elsewhere. The engineering and textile industries have contracted sharply over recent decades, and few factories of any kind now use coal directly as an energy source. Most remaining steel works are still in their traditional locations, even though most iron ore is now imported (Fig. 3.7). The modern industry most directly tied to coalfields is electric power generation (Fig. 3.7). However, coal-fired power stations themselves are now threatened by cheaper natural gas and by concern for the atmospheric effluents from burning coal (Ch. 17).

Geological resources other than coal have influenced British geography (Fig. 3.7). The early centres of the chemical industry, on Merseyside and Teesside, are both close to raw materials provided by Permo-Triassic salt deposits (Fig. 3.7). Their coastal sites allowed the import of other materials such as fats and

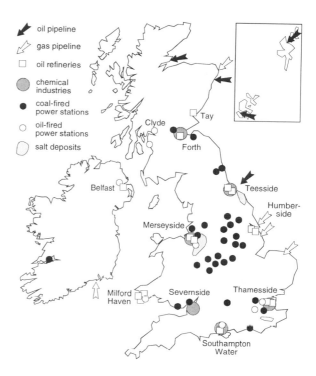

Figure 3.7 The location of some geologically influenced industries.

vegetable oils, and both areas had local coal, their former energy source. With the development of **petrochemicals**, based on natural gas and **naptha** from refined oil, the Merseyside and Teesside industries could use feedstock from the oil refineries established on the same estuaries. Petrochemical works grew up near other major refineries on the Firth of Forth, Thamesside and Southampton Water.

Petroleum supply and refining has its own peculiar geography (Fig. 3.7). North Sea oil and gas are piped ashore at East Coast terminals, some with associated oil refineries. But oil from small and remote fields is shipped to such terminals as Milford Haven, Southampton Water and Thamesside, built to import oil before North Sea resources were found. It is a nice historical contrast that the ease of transporting oil and gas by pipeline has meant that oil terminals and refineries have not attracted industry and population around them as coal did in the 19th century. Future geographical shifts await a new stimulus, perhaps renewable energy sources?

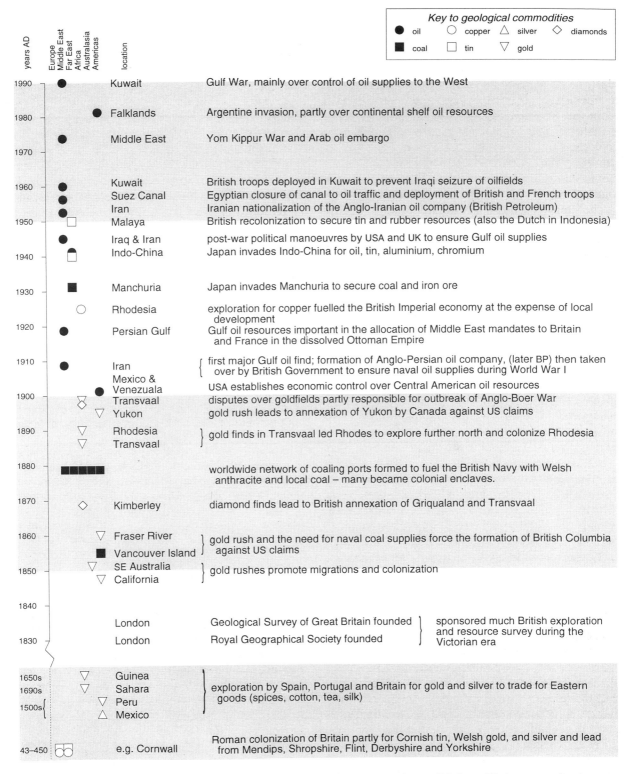

Figure 3.8 Some geologically influenced historical events, concentrating on Britain and its former empire.

3.4 Geology and historical geography

This chapter has shown how the present human geography of the British Isles results from modern factors overprinted on legacies from the past, some of them from the geological past. Some historical events elsewhere in the world have also been influenced by geological factors, particularly the uneven global distribution of resources. A time-chart of such events, emphasizing British and Colonial history, is a catalogue of migration, colonization and war (Fig. 3.8).

The Roman occupation is an early example of geological influence on British history. The first conquests subjugated southeast England as far as the defensive line of the Jurassic limestone scarps in the Cotswolds and the Midlands. Later advances followed precisely the three salients where the Mesozoic lowlands extend out from the Midlands; northwards through Yorkshire to the Tees Estuary, northwestwards over the Cheshire plain to the Dee Estuary, and westwards across the Severn into South Wales. The resulting Roman Civil District broadly coincided with Mesozoic Britain. Beyond this was established the Military District of the Palaeozoic and Precambrian uplands. This area was mainly under Celtic control, but was planted with forts and linking roads partly to exploit its coal and, in particular, its metallic mineral wealth.

In a world perspective too, metallic minerals were the main geological catalyst to events until the 20th century (Fig. 3.8). Because base metals are common in many regions, it was exploration for the rare sources of gold and silver that had most international repercussions. The "gold rushes" of the 19th century were particularly strong agents of migration and colonization, supplemented in southern Africa by the discovery of diamonds.

The energy needs of most pre-industrial societies were met by wood, wind and water power. Fossil fuels were exploited only from exposed coal seams, or from natural oil or gas seeps. Coal led the way through the industrial revolution into the 20th century preoccupation with oil supplies. This shift of concern from minerals to fuels moved the geographical focus of disputes over resources from Africa, Australia and the Americas to the Middle and Far East (Fig. 3.8).

The list of resource-led international disputes is set to lengthen. Increasing global demand for remaining oil resources, heavily concentrated in the Middle East, promises further conflict. Natural gas may join oil as a source of disagreement, particularly over the abundant Siberian resources. If the threat of global warming were to be faced seriously, fossil fuels would be substituted by renewable or, less advisedly, nuclear energy. The nuclear option would place a premium on uranium supplies and on safe sites to bury radioactive waste. Finally, more fundamental Earth resources are now proving internationally contentious; water, some of it in the subsurface, and air, polluted mainly by emissions from the use of fossil fuels. The relevance of geology to human political geography looks set to continue.

Further reading

Aalen, F. H. A. 1978. *Man and the landscape in Ireland.* London: Academic Press. [A thorough analysis of the history of human settlement in the Irish landscape.]

Brunskill, R. W. 1987. *Illustrated handbook of vernacular architecture.* 3rd edn. London: Faber & Faber. [An excellent guide to buildings in England and Wales, including distribution maps of materials.]

Danaher, K. 1978. *Ireland's vernacular architecture,* 2nd edn. Cork: Mercier Press. [A well illustrated introduction to traditional buildings in Ireland.]

Dury, G. H. 1973. *The British Isles,* 5th edn. London: Heinemann. [A dated but still useful introduction.]

Irish National Committee for Geography 1979. *Atlas of Ireland.* Dublin: Royal Irish Academy. [An excellent compilation of data on physical and human geography.]

Johnston, R. J. & V. Gardiner (eds) 1991. *The changing geography of the British Isles,* 2nd edn. London: Routledge. [A review of changes in the geography of Britain over the last decade or two.]

Naismith, R. J. 1985. *Buildings of the Scottish countryside.* London: Gollancz. [One of the few comprehensive works on Scottish vernacular architecture.]

Penoyre, J. & J. Penoyre 1978. *Houses in the landscape.* London: Faber & Faber. [A well illustrated regional study of vernacular buildings in England and Wales.]

Watson, J. W. & J. B. Sissons (eds) 1964. *The British Isles: a systematic geography.* London: Nelson. [A standard and reliable work, particularly on the history of geographical patterns in both Ireland and Britain.]

Yergin, D. 1991. *The prize: the epic quest for oil, money and power.* New York. Simon & Schuster. [The comprehensive story of the influence of oil on world affairs.]

Chapter Four

GEOLOGICAL HAZARDS

4.1 Hazard type and assessment

Previous chapters have outlined how human activity in the British Isles is influenced by the region's geology, the result of Earth processes over hundreds of millions of years. These processes are continuing, some imperceptibly slowly, some rapidly enough to affect and threaten life. Landslides, subsidence, earthquakes, coastal erosion – these are some **geological hazards**. Together with **biological**, **meteorological** and **hydrological** hazards, they form a spectrum of natural threats (Fig. 4.1). This chapter will describe hazards involving the solid Earth, but including related hydrological processes such as floods and coastal erosion. It will concentrate, though not exclusively, on hazards relevant to the British Isles.

Technological hazards are due to human activity rather than natural processes (Fig. 4.1). The distinction is blurred, because many natural hazards can be triggered or accelerated by human actions. Mining causes ground subsidence, poorly sited buildings can

induce landslides, and the emission of greenhouse gases threatens climate change and more storms. Such specifically **human impacts** on the natural environment are the subject of Part IV.

Hazardous processes are not necessarily unusual. Indeed, some are regarded as **resources** when they operate within their normal intensity range (Fig 4.2). For example, wind, tides, river flow and volcanic heat are all valuable sources of energy (Ch. 12). Only when their intensity exceeds a **damage threshold** do they become **hazards**, or "malevolent resources". Some natural agents, such as rainfall and temperature, can be hazardous outside both a lower and an upper threshold. Many geological processes, such as earthquakes, landslides and coastal erosion, just have an upper damage threshold.

What are the globally important hazards? Floods account for more disasters than any other hazard, followed by cyclones and hurricanes, earthquakes and tornadoes (Fig. 4.3a). However, earthquakes and cyclones cause the most deaths in each event.

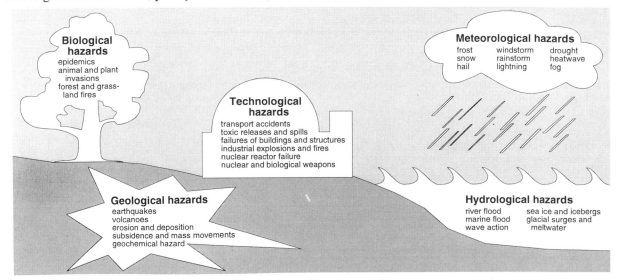

Figure 4.1   Geological hazards in the context of other natural and human-induced hazards.

Natural disasters are unevenly distributed across the world. Nearly 40% occur in Asia and another 33% in North America (Fig. 4.3b). More uneven still is the resulting loss of life, with over 85% of disaster deaths occurring in Asia. This imbalance shows that the **risk** to life from a natural process depends on the level both of the **hazard** and of the **vulnerability** of the population. The much lower death toll in North America than in Asia, in response to a comparable hazard frequency, is due to lower population densities, more strongly engineered structures, better emergency services, and other advantages of a wealthy society.

Worldwide, the annual risk of being killed in a natural disaster is around 1 in 10 000 (Fig. 4.4b). The relatively quiet geology and climate of the British Isles lower this risk substantially, to about 1 in 5 million for storms and floods, for instance (Fig. 4.4a). Radon gas (Ch. 4.8) is estimated to be the region's most potent natural risk at about 1 in 50 000, still less than the threats from air pollution and road accidents. For comparison, the annual mortality from all causes in the British Isles varies with age from 1 in 1 000 to over 1 in 10 (Fig. 4.4a). The prominence given to natural hazards in the industrialized world is due more to the threat to property and the workings of society than to life. Natural hazards probably cost the global economy between $50 billion and $100 billion each year during the 1980s. Insurance losses from one windstorm in Britain in October 1987 amounted to at least £1 billion, and from the Californian earthquake of October 1989 to about $1.5 billion. In the wealthy world these numbers count as much as death tolls.

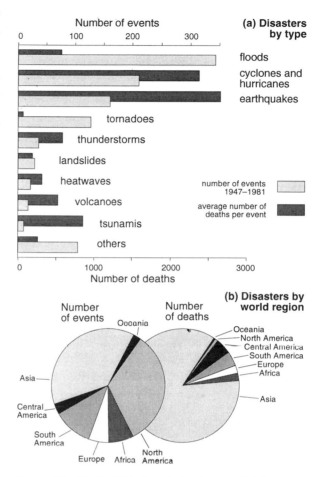

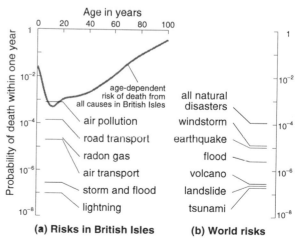

Figure 4.3 World distribution of disasters 1947–81.

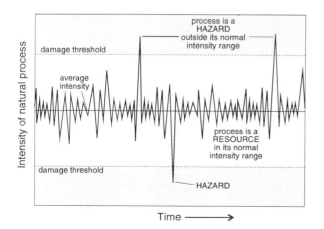

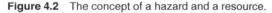

Figure 4.2 The concept of a hazard and a resource.

Figure 4.4 Some risk levels in Britain and worldwide.

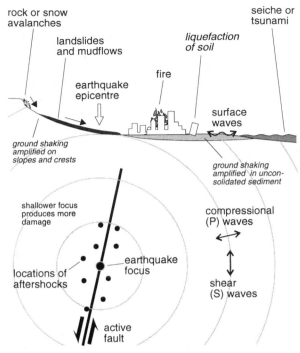

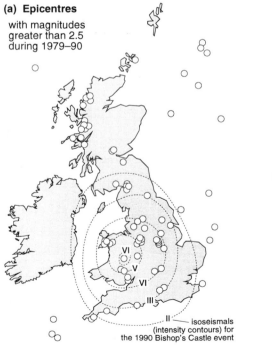

Figure 4.5 Mechanism and hazards of earthquakes.

4.2 Earthquakes

The British Isles, presently remote from the active plate boundaries where earthquakes are concentrated, suffers few damaging shocks. However, the risk is high enough to influence the design of sensitive structures such as dams and nuclear power stations.

Earthquakes result from rapid slip along **faults** (Fig. 4.5). The **seismic energy** from the slip travels out from the earthquake **focus** as both **compressional** or **P-waves**, vibrating along the direction of travel, and as **shear** or **S-waves**, vibrating normal to the travel direction. **Surface waves** are set up when the energy pulse reaches the ground surface.

The **ground shaking** due to the passage of these various waves causes the primary damage to structures. Damage is more severe if the earthquake is large, if its focus is shallow, or if structures are built on soft sediment that amplifies the wave motion. Secondary hazards cause more damage and loss of life. Fires are started from ruptured fuel pipes and power lines. Unconsolidated wet sediments undergo **liquefaction**, weakening foundations and toppling build-

(a) Epicentres

with magnitudes greater than 2.5 during 1979–90

(b) Intensity scale

I   Not felt by people. Recorded bu seismographs only.

II   Felt by some people at rest on upper floors.

III   Felt noticeably indoors by some people. Hanging objects may swing slightly.

IV   Felt indoors by many, outdoors by a few. Some sleepers awakened. Windows, dishes and doors rattle. Liquids disturbed. Standing cars rock.

V   Felt by nearly all. Many sleepers awakened. Open doors and windows swing. Unstable objects overturned. Some plaster cracks in weak buildings.

VI   Felt by all. Many frightened and run outdoors. Dishes break. Furniture moved. Plaster falls in sound buildings. Chimneys of weak buildings damaged.

VII   Difficult to stand. Most run outdoors. Noticed in moving cars. Damage to plaster, brickwork and tiles even in sound buildings. Waves on ponds.

VIII–XII   More severe effects, not generally experienced in earthquakes in Britain.

(c) Magnitude scale

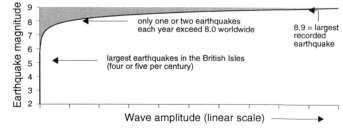

Figure 4.6 Distribution, magnitude and intensity of earthquakes in the British Isles.

ings. **Landslides**, **mudflows** and **avalanches** may be triggered on slopes (Ch. 4.4). Catastrophic surges of water may be set up in the sea (**tsunamis**) or in lakes (**seiches**).

One way of measuring the size of an earthquake is to assess the damage and human responses that it provoked. The resulting scale of earthquake **intensity** runs from I to XII (Fig. 4.6b). The intensity decreases outwards from the **epicentre** of an earthquake, directly above its focus. Intensity contours (or **isoseismals**) for a large British earthquake show an epicentral intensity of VI (Fig. 4.6a). The most damaging British shocks have had a maximum intensity of VIII.

Instrumental measuring of ground vibrations by **seismometers** yields more quantitative data, calibrated on a scale of **earthquake magnitude** (Fig. 4.6c). An increment of 1 on the magnitude scale represents a tenfold increase in the amplitude of earthquake waves, and a thirtyfold increase in the total energy released. The largest earthquakes in Britain have had magnitudes between 5.0 and 5.5, releasing insignificant energy by comparison with globally important shocks. Most British earthquakes are much smaller still; on average only about seven exceed magnitude 2.5 each year. Their spatial pattern is poorly understood, particularly why Ireland is seismically quiet (Fig. 4.6a).

4.3 Volcanoes

If the earthquake hazard in the British Isles seems small, the volcanic risk might seem negligible, so distant is any active volcano. In fact, the dust and gas injected into the **stratosphere**, the atmosphere above about 15km, from very large explosive eruptions has global effects (Fig. 4.7). Temporary cooling of the global climate and degradation of the radiation shield of stratospheric ozone are detailed in Chapter 17. The second of these is a global threat at least as serious as conventional volcanic hazards.

Primary hazards reflect the type of host volcano (Fig. 4.7). **Andesitic volcanoes**, typical of subducting plate boundaries, tend to produce more viscous **magma** that sticks in a volcanic vent before erupting explosively. A column of ejected material either blasts upwards, to deposit **airfall tephra**, or rolls down the volcano slope as a hot **pyroclastic flow**. **Basaltic volcanoes**, typical of rifting plate boundaries, produce more fluid magma that forms steady **lava flows**. Both types of volcano emit **volcanic gases**, and can trigger secondary hazards such as volcanic mudflows (termed **lahars**), **seiches** and **tsunamis**. Many volcanic eruptions have **precursor events** that might be used for hazard prediction (Fig. 4.7).

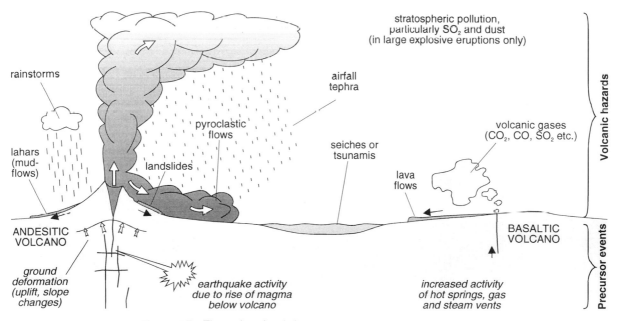

Figure 4.7 The main volcanic hazards and their precursory events.

(a) Types of mass movement

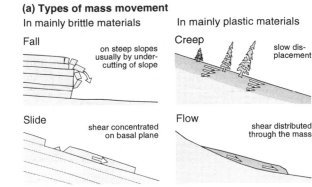

(b) Causes of mass movement

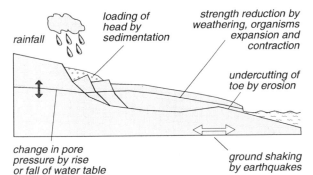

Figure 4.8 The main type of mass movement hazard.

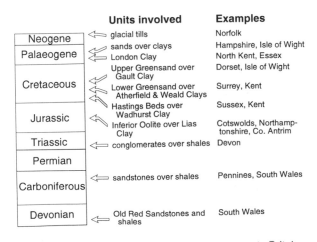

Figure 4.9 Main stratigraphic levels of landslides in Britain.

4.4 Landslides

The term **landslide** is used in a general sense for the spectrum of downslope **mass movements** that affect surface rock and soil. Landslides are classified by their shape and their flow behaviour (or **rheology**): are they made up of cemented rock that deforms only by **brittle fracture**, or of soil or unconsolidated rock that can deform by **plastic flow** (Fig. 4.8a)?

Brittle materials typically develop **slides**, in the strict sense, where a rigid block or slab moves down slope on a discrete slide or **shear plane**. This plane often follows a pre-existing bedding plane in sedimentary rocks or a joint or fault. Rockslides are usually rapid, travelling at many tens of kilometres an hour. A **fall** is the limiting case where blocks slide or topple off a nearly vertical slope. Repeated falls can occur only where erosion removes the fallen debris to maintain the slope, for example on a coastal cliff or river bank.

Plastic materials, particularly soil, can suffer steady or seasonal **creep** at rates of about 1 to 100 mm/year. More catastrophic **flow** events can involve hundreds of tonnes of material moving at up to 100 km/hour. The shear displacements are distributed through the full depth of the flow rather than being confined to its basal plane. Flows can affect weak mudrocks as well as soil. They can also occur in sands or even rock debris, particularly when waterlogged.

Slumps are rotational slope failures above a circular shear surface, spoon-shaped in three dimensions (Fig. 4.8a). Slumps are typical of materials that lack bedding or fracture planes dipping parallel to the slope. They can move partly as rigid blocks along discrete shear planes. However, plastic flow accommodates the irregular shape changes, particularly at the downslope **toe** of the moving mass. Slumps tend to move less far and less rapidly than planar slides and flows.

Mass movements are caused by natural factors (Fig. 4.8b) and by some forms of human interference that are detailed later (Ch. 15). Changes in the geometry of the slope can make it unstable, particularly erosion on its lower part and deposition on its upper part. A rise in the **water table**, the level of water in the pores of the rock or soil, can increase the plasticity of the mass and reduce the frictional resistance along potential failure planes. Heavy or prolonged rainfall causes a rise in water level and can trigger a catastrophic land-slide. The slope material can also be weakened by the slower processes of **weathering**; chemical alteration of minerals, physical breakdown by freezing and thawing, and biological disruption by soil organisms and plant roots. Less important in the British Isles, but a potent trigger for landslides worldwide, is shaking by earthquakes. This can break the cohesion across fractures and liquefy waterlogged sands and muds by disturbing their grain structure. Volcanic eruptions also promote landslides, termed **lahars**, triggered variously by rapid accumulation of volcanic debris, by release of crater lake water, by rainfall, or by earth-quakes.

Most landslides in the British Isles develop in weak shales, especially where they are overlain by more dense and strong sandstones or limestones. Landslides are therefore concentrated along the outcrops of spe-cific stratigraphic units (Fig. 4.9). The most common hosts are Mesozoic and Cenozoic clays in the south and east of Britain. These clay-controlled landslides occur extensively inland, for example along the north-west-facing escarpments of the Jurassic limestone belt, but are an important process in the erosion of parts of the south coast of England.

Some of the largest landslides of inland Britain oc-cur in the Upper Carboniferous of northern England, where thick units of sandstone are interbedded with shales. An example on the eastern slope of Mam Tor, in the southern Pennines, is a weakly rotational slump with its basal shear in weak shales (Fig. 4.10). The head of the slump is loaded by an active scree from the overlying sandstones that are exposed in the main scarp. The historical retreat of this scarp is demon-strated by its having truncated the rampart of an Iron Age hill fort. Continuing creep of the main landslide, at up to 0.6 m/year has caused the abandonment of the major road that used to cross it.

Many small landslides have developed in Quater-nary deposits, particularly in the upland areas of the north and west of the British Isles. Glacial sediments that have suffered slow downslope creep form the deposit termed **head**. Peat can become so waterlogged after heavy rain that it bursts through its confining skin of vegetation and moves down slope as a rapid flow. These **bog bursts** are most frequent in high rainfall areas, for instance in Ireland (Fig. 4.11).

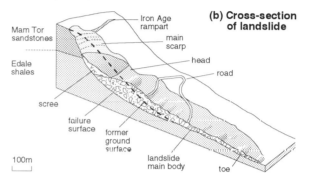

Figure 4.10 The Mam Tor landslide, north Derbyshire. Part b is reproduced from *Geoscientist* (Cripps & Hind 1992) with the permission of the Geological Society Publishing House

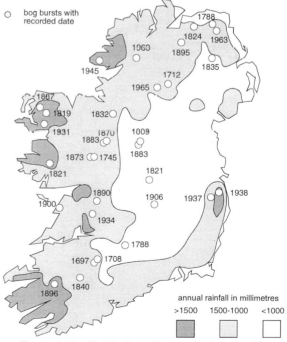

Figure 4.11 Distribution of bog bursts in Ireland.

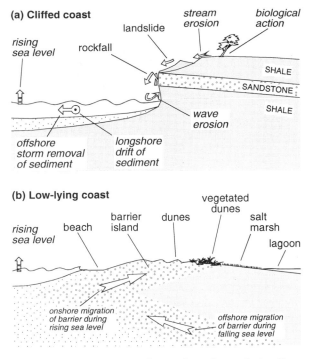

(a) Cliffed coast

(b) Low-lying coast

Figure 4.12 Processes of coastal erosion and migration.

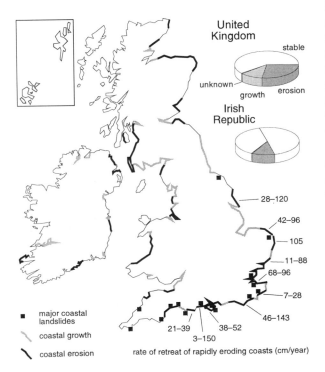

major coastal landslides

coastal growth

coastal erosion

rate of retreat of rapidly eroding coasts (cm/year)

United Kingdom

Irish Republic

Figure 4.13 Coastal instability in the British Isles.

4.5 Coastal erosion

Coastlines are geologically unstable features. New land develops as sediment is deposited and is stabilized by growth of plants. Such land is a valuable agricultural resource, such as in the Fens of eastern England. However, **coastal erosion** can be a major hazard. Although average erosion rates may be low enough to predict and accommodate, coastal retreat is punctuated by catastrophic events such as landslides or floods, which threaten life and property.

Cliffed coasts, developed on resistant rocks, erode most directly by wave action at their base (Fig. 4.12a). However, the continual steepening and undercutting cause **rockfalls** from strong rock units and **landslides** in weaker units such as mudstones and shales. Disintegration of cliffs is helped by weathering agents such as tree roots, and by the erosion in cliff-top streams. The debris washed from the cliff base may be transported offshore during large storms, or alongshore by fair-weather inshore currents.

Low-lying coasts commonly have a raised **barrier** near the shoreline, backed by a low-lying **salt marsh** or **lagoon** (Fig. 4.12b). The sediment for the barrier is supplied by wave action, then blown inland to form **dunes** to be stabilized by vegetation. The barrier can migrate seaward, if the supply of sediment exceeds its removal during storms. However when this sedimentary balance is reversed, particularly during periods of rising sea level, the barrier migrates inland and the shoreline retreats. Regional subsidence may preserve the buried sedimentary record of both the advance and the retreat phases (Fig 4.12b).

There are important human impacts on coastal processes, both through attempts to control erosion by engineered structures (Ch. 15) and due to the acceleration of sea-level rise threatened by global warming (Ch. 17).

Coastal erosion exceeds coastal growth in the British Isles as a whole (Fig. 4.13). However, eroding coasts are concentrated in the south and east of England, where sea level is rising relative to land and where soft Mesozoic and Cenozoic rocks predominate. This area hosts most of the large coastal landslides in Britain and most of the coasts with very high retreat rates, exceeding a metre a year along parts of Yorkshire, East Anglia and the South Coast.

4.6 Floods

Floods are a **hydrological hazard** resulting from high inland rainfall or snowmelt, or from high tides and storms on the coast (Fig. 4.14). Geological data can help to predict flood-prone areas and to design measures to minimize the flood hazard (Ch. 15).

River **flood plains** are the most hazardous inland areas (Fig. 4.14). The risk is high if the drainage basin is prone to concentrated rainfall and is underlain by impermeable rocks. For instance, the basins of the Severn, Wye, and the South Wales rivers drain the wet Welsh uplands over Ordovician to Carboniferous rocks and impermeable glacial till (Fig. 4.15). Over 7000 urban properties are at risk, although water-supply reservoirs and local flood-prevention schemes are reducing this number. Reservoirs themselves present an extra, if small, hazard. Major floods are occasionally caused by failure of dams or by large landslides displacing water from a reservoir.

Sea floods occur along low-lying coasts (Ch. 4.5). A large storm during a high tide can overtop or breach a natural barrier or an engineered sea defence, flooding land close to or below sea level. The worst flood in Britain, in January 1953, killed over 300 people and caused £50 million of damage (Fig. 4.16). Its effects were amplified in estuaries where high tidal inflow met already high river water (Figs 4.14 & 16). The relative rise in sea level in southeast Britain of around 0.5 m per century is increasing the coastal flood risk, especially if this rise is being enhanced by the impact of global warming (Ch. 17).

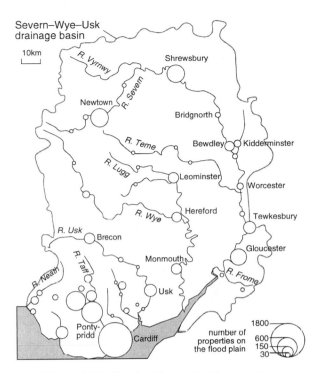

Figure 4.15 Flood risk in the Welsh borderland.

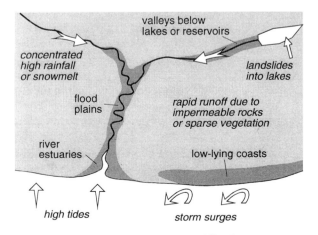

Figure 4.14 Flood-prone areas and flood processes.

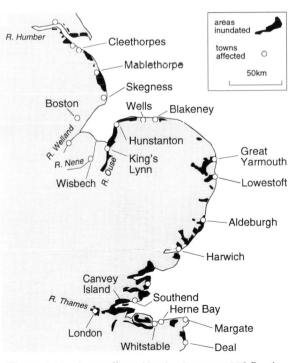

Figure 4.16 Areas affected by the January 1953 floods.

4.7 Subsidence

Subsidence is the vertical lowering of the ground surface. It presents a major natural hazard to property and engineered structures, and can result in loss of life when it is rapid and localized. Human activities can trigger or enhance subsidence (Part IV), but there are

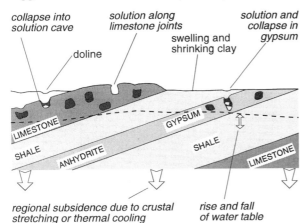

Figure 4.17 Some mechanisms of ground subsidence.

three main natural mechanisms (Fig. 4.17).

Tectonic subsidence is a slow widespread process. It affects the North Sea and southeast England, contributing to relative sea-level rise and coastal retreat.

Swelling and shrinking clays are able to absorb and release water from their mineral structure. Soils rich in these clay minerals expand when wetted and contract when dried, causing structural damage to foundations. A series of dry summers in southeast England during 1989–91 highlighted clay soils as a serious factor in land-use. The distribution of these soils is shown in this context in Figure 6.7.

Solution subsidence involves the collapse of overlying rock into cavities dissolved out of limestones or evaporites (Fig. 4.17). Susceptible units are widespread in the British Isles (Fig. 4.18a), and have been through several episodes of dissolution or **karstification**. Measured solution rates of between 25 and 100 m³/km²/yr show that significant dissolution is happening at the present day. An example of active salt solution in North Yorkshire shows arrays of collapse hollows or **dolines** (Fig. 4.18b).

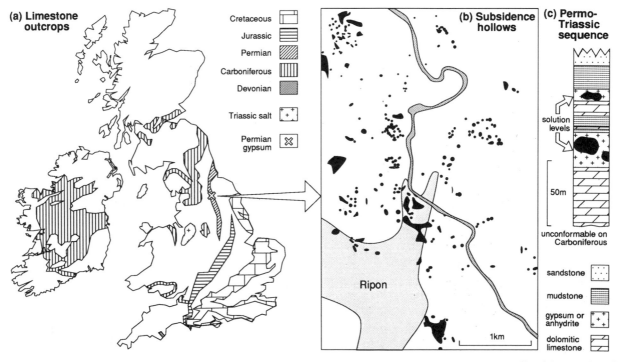

Figure 4.18 Limestones and evaporites prone to solution subsidence, with an example from the Ripon area. Part b is reproduced from *Special publication 22* (Cooper 1986) with the permission of the Geological Society Publishing House.

4.8 Geomedical hazards

Most natural hazards threaten life and property in direct, physical ways. More insidious, but probably more dangerous in the British Isles, are the chemically transmitted health hazards posed by rocks that have an excess or deficiency of life-influencing elements. The natural pathways for these geochemical anomalies are through water, soil and plants, and through animals (Fig. 4.19a). These hazards from ingestion are supplemented by those from inhaling air and from the many polluting activities of human beings (Part IV).

The **essential macronutrients** for life are those required in relatively large amounts, more than 100 mg per day in humans (Fig. 4.19a). **Essential micronutrients** are needed in daily quantities of less than a few milligrams. **Toxic elements** have no beneficial use but are harmful in excess. Geochemical surveys of rocks and soils highlight anomalies in the availability of these elements. Establishing firm correlations with medical symptoms is more difficult, but some of the more certain links can be listed (Fig. 4.19b). The range of human and animal tolerance of many elements is narrow, so that both a deficiency and an excess can have adverse effects.

A rather different geomedical threat is posed by **radon**, a gas derived by radioactive decay of uranium and thorium in rocks. Radon is itself radioactive and inhaling it increases the risk of lung cancer. Most radon from rocks escapes through pores and cracks into the atmosphere, where it is diluted to safe levels. It can reach dangerous concentrations when trapped in mines, caves and unventilated buildings.

Regional surveys have highlighted the areas in Britain with a high radon risk (Fig. 4.20). These areas mostly contain rocks with a high uranium content; **granite** in southwest England and parts of Scotland, **phosphate rock** such as in Northamptonshire, and **black shale** in Wales and the Carboniferous coal-bearing sequences. High radon values occur over **limestone** in Derbyshire due to its high fracture permeability. More detailed house-by-house surveys are needed in high-risk areas, because radon levels depend strongly on local geological details of lithology and fracture locations. Affected buildings and mines simply require better ventilation to reduce the radon risk.

(a) Natural trace-element pathways

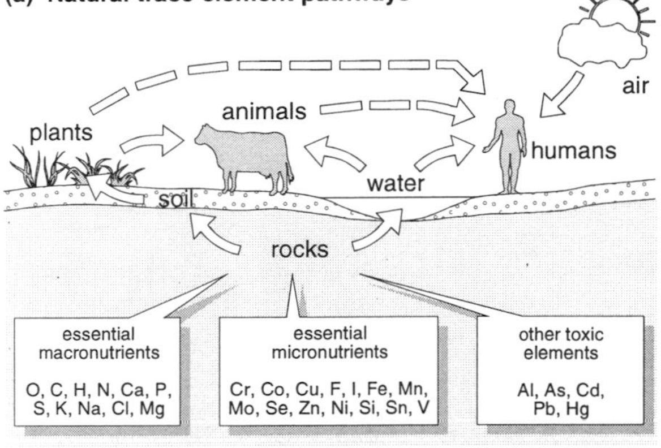

essential macronutrients	essential micronutrients	other toxic elements
O, C, H, N, Ca, P, S, K, Na, Cl, Mg	Cr, Co, Cu, F, I, Fe, Mn, Mo, Se, Zn, Ni, Si, Sn, V	Al, As, Cd, Pb, Hg

(b) Medical effects of some elements

Element		Effects of deficiency	Effects of excess
Na	sodium		high blood pressure
Cr	chromium	glucose intolerance	dementia, cancer
Cu	copper	anaemia, bone demineralization, high cholesterol	blood disease
F	fluoride	tooth decay, bone loss	excess bone growth, ligament calcification
I	iodine	goitre	goitre
Fe	iron	anaemia	
Se	selenium	heart disease	tooth decay, birth defects, neurological disorders
Zn	zinc	immune deficiency, skin disease	anaemia

Figure 4.19 Some pathways and effects of trace elements.

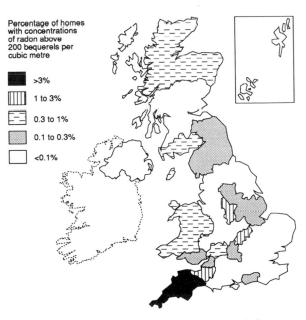

Percentage of homes with concentrations of radon above 200 bequerels per cubic metre

- >3%
- 1 to 3%
- 0.3 to 1%
- 0.1 to 0.3%
- <0.1%

Figure 4.20 The risk from radon gas in Britain.

4.9 Future impacts of natural hazards

Most assessments of worldwide hazard show a trend of disaster numbers and impacts increasing through time (Fig. 4.21b). One reason for this trend is improved global monitoring and reporting. For instance, the annual number of volcanoes reported as active has doubled over the past century, although it decreased during the two world wars. This variation does not reflect a real change in volcanic activity. Rather, each hazardous event is having a greater impact on life and property. The number of major disasters increased fivefold from the 1960s to the 1980s (Fig. 4.21b), with a corresponding increase in the number of affected people. The economic losses from these major disasters increased by over three times in the same period (Fig. 4.21a). Predictably, economic losses are greatest in developed countries and loss of life greatest in the developing world.

These disaster trends are due to the increasing **vulnerability** of populations and property to hazard events. **Population growth** forces development of high-risk land such as flood plains and seismic zones. Increasing **urbanization** can concentrate people in these same areas, usually in poorly engineered housing. **Economic growth** increases the complexity and value of infrastructure and property. Growing **interdependence** of populations for food, water and industrial materials spreads the regional impact of an otherwise local disaster. These pressures make the assessment of and preparedness for natural hazards more important than ever before.

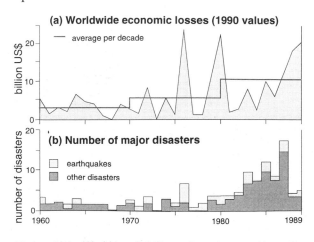

Figure 4.21 World trends in hazard numbers and impact.

Further reading

Allison, R. J. 1990. *Landslides of the Dorset coast*. London: British Geomorphological Research Group. [A field guide to one of England's main landslide-prone areas.]

Alexander, D. 1993. *Natural disasters*. London: UCL Press. [A modern scientific review of all natural hazards.]

Bolt, B. A., W. L. Horn, G. A. Macdonald, R. F. Scott 1977. *Geological hazards*. New York: Springer. [Descriptive coverage of hazards based on North American examples.]

Coates, D. R. 1981. *Environmental geology*. New York: John Wiley. [Part 3 covers geological hazards, and is particularly informative on landslides and floods.]

Department of the Environment 1992. *The UK environment*. London: HMSO [Chapter 9 provides data on coastal erosion and flooding and Chapters 13 and 14 include material on radon and geochemical hazards.]

Griggs, G. B. & J. A. Gilchrist 1983. *Geologic hazards, resources, and environmental planning*. Belmont, California: Wadsworth. [Thorough, well illustrated guidebook to geological hazards, their prediction and mitigation, based on North American examples.]

Lumsden, G. I. (ed.) 1992. *Geology and the environment in Western Europe*. Oxford: Oxford University Press. [Chapter 4 is a useful assessment of geological hazards in Europe.]

McCall, G. J. H., D. J. C. Laming, S. C. Scott (eds) 1992. *Geohazards: natural and man-made*. London: Chapman & Hall. [A useful collection of papers on geological hazards in a global perspective.]

Chester, D. 1993. *Volcanoes and society*. London: Edward Arnold. [An excellent survey of volcanic processes and hazards.]

Perry, A. H. 1981. *Environmental hazards in the British Isles*. London: Allen & Unwin. [Uses British and Irish examples, but concentrates on atmospheric and hydrological hazards.]

Smith, K. 1992. *Environmental hazards: assessing risk and reducing disaster*. London: Routledge. [An excellent introduction to risk assessment, with sections on the main hazard types.]

van Rose, S. 1983. *Earthquakes*. London: HMSO. [Short, authoritative and well-illustrated.]

Thornton, I. (ed.) 1983. *Applied environmental geochemistry*. London: Academic Press. [Collection of articles detailing principles and some case studies of the geomedical hazard.]

Wozniak, S. J. 1992. *Handbook of radon*. Abbots Langley, England: Wozniak. [An authoritative review of the radon risk in Britain.]

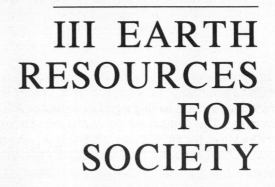

III EARTH RESOURCES FOR SOCIETY

Headgear at a disused coal mine in the North Wales coalfield: a victim of the declining market for British coal in the 1990s.

Resources are the sustenance of human society: land to live on, air to breathe, plants and animals to eat, and rainwater to drink. Less vital, but now essential to developed societies, are resources from below the Earth's surface. These **geological resources** are the subject of Part III. Some general concepts of Earth resources are introduced in Chapter 5, particularly the problem of estimating their available reserves. The geological aspects of **land** as a resource are covered in Chapter 6, and of **water**, particularly ground-water, in Chapter 7. Chapter 8 describes the rocks that are used as construction materials for built structures. Minerals are covered in Chapter 9, both the **metallic minerals** and the increasingly important **industrial minerals**. Chapters 10 and 11 introduce fossil fuels, the **coal, peat, oil** and **gas** that satisfy most of the world's massive requirement for energy. Finally, Chapter 12 looks at sources of alternative energy, including uranium and geothermal energy.

Part III concentrates on the formation and availability of Earth resources. The impacts of using and extracting them are the subject of Part IV.

Chapter Five

RESOURCES

5.1 Types of geological resources

Coal, oil and gas, minerals and rocks: these are typical geological resources. They are usually considered to be **non-renewable**, with society progressively depleting a fixed stock of each commodity. By contrast, water or air are usually termed **renewable**, because natural processes replenish and recondition the stock as it is used.

However, renewability is not a simple measure. All resources are renewable on some timescale. For instance, oil and gas are forming now in the world's sedimentary basins, and mineral deposits beneath active volcanoes. On the other hand, water from rainfall

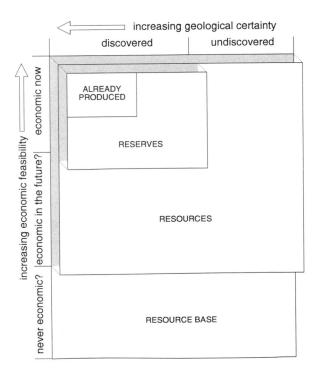

Figure 5.1 The McKelvey resource classification scheme.

may not be adequate to refill reservoirs or rock aquifers. A more helpful measure is the **sustainability** of a resource; whether or not its rate of use exceeds its rate of renewal. Most geological resources are **unsustainable**, because their formation processes are very slow on a human timescale. Oil is being used at least a million times faster than it is being recreated. Water and land are potentially **sustainable resources**, but only if managed correctly.

Unsustainable resources make up most of the subsurface materials that geologists are called on to assess. A widely used method for assessing their availability is the **McKelvey scheme** (Fig. 5.1). In this, the **resource base** of a commodity is the total amount that exists on Earth. For most commodities this amount is of no practical interest, because much of it could never be economically exploited. The **resources** represent the part of the resource base that might conceivably be economic in the future. Within this amount, only the **reserves** are both economic now and identified with some geological certainty. A final fraction, previously part of the reserves, has been **already produced** and used by society.

The criteria of the McKelvey scheme mean that estimates of reserves and resources vary with changes in economic conditions and geological knowledge (Fig. 5.2a). For instance, reserves appear to increase if the price of a commodity rises, making it attractive to exploit lower-grade and less accessible resources. Conversely, increased costs of extraction and processing will lower the assessed reserves. Estimates of resources are dependent on geological assumptions about their formation and occurrence. Refinement of these geological models can either increase or decrease resource estimates.

Estimates are further complicated because changes in the reserves can themselves affect economic and social activity, forming feedback loops that slow the potential changes. So, apparent shortages in reserves raise the price of a commodity and therefore the pace

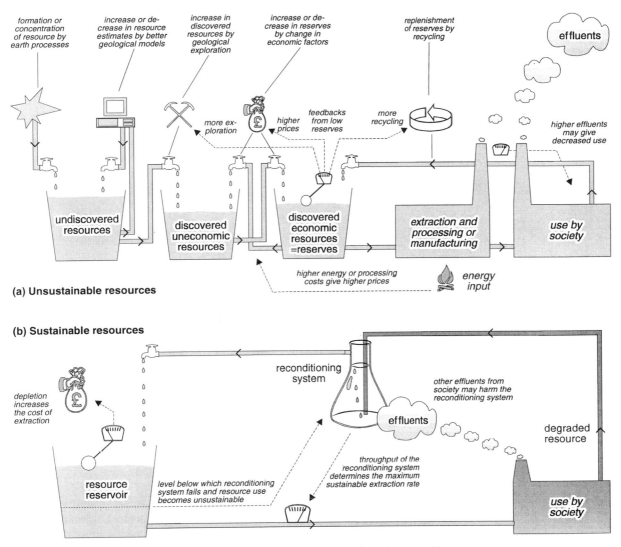

Figure 5.2 Controls on the use of unsustainable and sustainable resources.

of geological exploration, both factors that tend to increase the reserves again. Low estimates of reserves also stimulate recycling of some commodities such as metals, slowing the rate of depletion of the natural stock. Finally, the use of some geological resources may be restrained not by the shortage of reserves but by the shortage of safe places to dump the effluents from their production and use (Chs 14, 16). The carbon dioxide derived from burning fossil fuels is the most serious example of this constraint (Ch. 17).

Sustainable resources are part of a cycle where the rate of use does not exceed the rate of natural replenishment (Fig. 5.2b). However, this balance involves quality as well as quantity. A geological resource such as groundwater is used and returned to nature in a dirtier, more degraded state than it was extracted. Natural reconditioning systems such as rivers, plants, evaporation and rainfall must be able to clean water fast enough. These systems are easily damaged by pollution, or even by the over-use of water itself. So excessive pumping can lower underground water levels, dry up natural springs, and starve the rivers which would have helped to purify the pumped water after use.

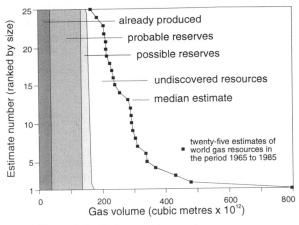

Figure 5.3 Estimates of natural gas resources.

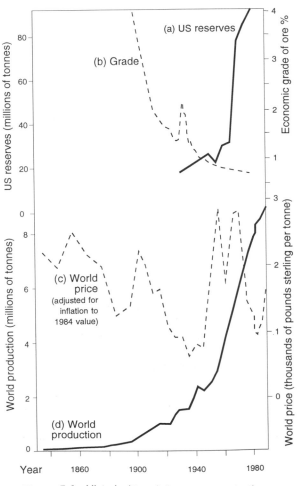

Figure 5.4 Historical trends in copper production.

5.2 Resource assessment

Estimates of the future availability of geological resources are subject to much uncertainty. One measure of this is to plot all recent assessments of a particular commodity, for example those for gas over a recent twenty-year period (Fig. 5.3). While there is agreement over estimates of gas reserves, there is a discrepancy of over four times in estimates of total resources. This variation is due to the different weighting given to geological factors in the assessment procedure; parameters such as the size of sedimentary basins, their depth of burial and the nature of their rock fill. A **median estimate** provides a consensus of current opinion, but is still prone to major revision as scientific understanding grows.

A common measure of the availability of a material is its **reserves:production ratio**: the number of years that reserves would last at the current rate of use. This number should not be interpreted literally as the lifetime of a commodity. The history of copper demonstrates the pitfalls of this simplistic view. World production of copper has risen sharply through time (Fig. 5.4d). The increasing call on world resources should logically have depleted copper reserves. In fact the estimated reserves in most producing countries have increased (Fig. 5.4a). This paradox is explained by the steady decrease in the grade of copper ore that it has been economic to mine (Fig. 5.4b). With each fall in grade, a new fraction of the resource is added to the reserves, a fraction that so far has been larger than the fraction just depleted.

A rise in demand for copper might also be expected to drive up its price. In fact the copper price has fluctuated around a constant value for over a century (Fig. 5.4c). This trend, common to many unsustainable resources, results from several feedbacks. A price rise stimulates exploration and the discovery of more reserves. It also promotes the substitution of other materials and the recycling of scrap metal. The result is a slowed rise in demand for copper ore from a now increased reserve, moderating or reversing the incipient rise in price.

An optimistic reading of trends such as those for copper (Fig. 5.4) is that resources will always be available to society. Growth in demand will be met from ever-increasing low-grade reserves, moderated only by the price that society is prepared to pay. This

Ricardian perspective on resources contrasts with the **Malthusian perspective** which reasons that any finite resource must eventually run out, especially if its exploitation grows exponentially through time.

The Ricardian view assumes that resources have a **unimodal** grade distribution (Fig. 5.5a), with society currently exploiting its high-quality tail. On this view, large reserves of lower-grade material are available at steadily increasing extraction costs. Although some common materials such as limestone or ironstone have a unimodal grade curve, many metals may have a **bimodal** distribution (Fig. 5.5b). Most of the metal is dispersed in low-grade silicate minerals with only a small fraction concentrated in economically attractive sulphides, oxides and carbonates. This mineralogical barrier coincides with a large jump in the costs of extraction, sharply reducing effective reserves and bringing closer the depletion envisaged in the Malthusian view.

The high costs of low-grade resources are those of land degradation (Ch. 14) and particularly of the energy needed to mine and concentrate low-quality metal ore, to pump up dispersed oil or to mine thin, deeply buried coal seams. The damaging atmospheric effects of the waste gases from this energy will prove to be a major new factor in costing the future use of all geological resources (Ch. 17).

Further reading

McLaren, D. J. & B. J. Skinner 1987. *Resources and world development*. Chichester: John Wiley. [Contains several informative chapters on resource assessment in general, as well as much material on individual commodities.]

Meadows, D. H., D. L. Meadows, J. Randers, W. W. Behrens 1972. *The limits to growth*. New York: Universe. [A Malthusian view of the future resource crisis, and the catalyst for a large later literature.]

Myers, N. (ed.) 1985. *The Gaia atlas of planet management*. London: Pan. [The first of many popular graphic compilations of data on world resources.]

Pearce, D. W. & R. K. Turner 1990. *Economics of natural resources and the environment*. New York: Harvester Wheatsheaf. [A detailed analysis of resource economics.]

Skinner, B. J. 1969. *Earth resources*. Englewood Cliffs, NJ: Prentice-Hall. [A useful introduction to geological resources.]

Simmons, I. G. 1991. *Earth, air and water: resources and environment in the late 20th century*. London: Edward Arnold. [A clear and valuable introduction to the changing rôle of resources in society.]

World Commission on Environment and Development 1987. *Our common future*. Oxford: Oxford University Press. [An influential analysis of prospects for global sustainable development.]

World Resources Institute 1992. *World resources 1992–93*. Oxford: Oxford University Press. [A biannual compilation of data and analysis on human and natural resources.]

Worldwatch Institute 1992 [and annually]. *State of the World 1992*. London: Earthscan. [An annual report on "progress towards a sustainable society", including readable analyses of natural resources.]

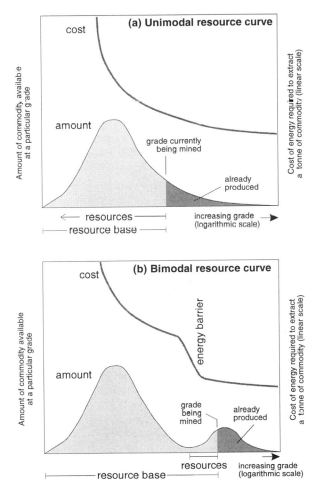

Figure 5.5 Two theoretical relationships between grade of a resource and its abundance, with resulting energy costs of extraction.

Chapter Six
LAND

6.1 Land as a resource

Land is a fundamental natural resource, but one of the most difficult to assess and use wisely. Land is naturally so variable and is used by society in so many different ways. For example, a river floodplain might be ideal for arable farming but too flood-prone to risk as an industrial site.

One approach to classifying land is to assess the present **land-use**. Usage categories for a temperate region such as the British Isles include crops, grass, rough grazing, forest and woodland, water and wetland, and urban and other built-over land (Ch. 6.3). However, the present use of land may not show its potential for future use. This assessment of **land capability** depends on observation of more fundamental factors (Fig. 6.1a). These include the solid and superficial geology, as well as landforms, vegetation, climate and hydrology. Soil type is particularly important in constraining agricultural use (Ch. 6.2). Landscape is a more subjective quality, although often crucial in planning land-use.

Many areas of land are at risk from natural hazards, most of which must be predicted rather than observed (Fig. 6.1b). These hazardous processes have already been described (Ch. 4). They limit the possible uses for land, not least because they may be triggered or accelerated by particular human activities. An example is the increased risk of soil erosion from some forms of agriculture (Ch. 15.6).

Land is not a renewable resource on a human time-scale. Natural processes act to both destroy and create land and to increase and decrease its quality. However, these are slow processes compared with human activities, many of which quickly **degrade** land for other uses (Fig. 6.2). Good planning of land-use restricts more severely the activities that degrade the land most – those highest on the list in Figure 6.2 – and encourages those agricultural and recreational uses that can preserve or even enhance land quality.

The various impacts of human activity on land use are detailed in Part IV.

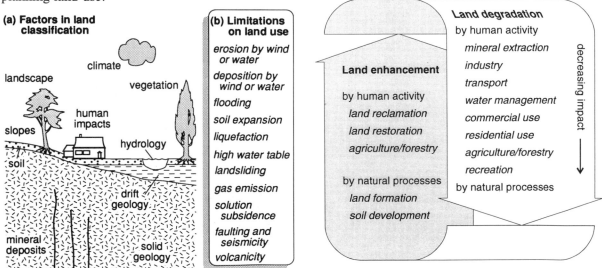

Figure 6.1 Land classification and assessment.

Figure 6.2 The cycle of land transformation.

6.2 Soil

Soil is the product of **weathering** of rocks at the earth's surface. **Mechanical weathering** breaks the rock down into smaller fragments by thermal expansion and contraction, or by freezing and thawing of water in cracks. **Chemical weathering** alters the composition of rock minerals by reaction with water or air. **Biological weathering** operates both by physical processes, such as tree roots wedging open cracks, and by organically produced chemicals. With time the thickness of weathered material increases, so long as it is not removed by **erosion**.

Three gradational zones can be recognized in many **soil profiles** (Fig. 6.3). In the top zone (A) rainwater moves downwards and dissolves soluble minerals, a process termed **leaching**. Organic matter is mixed into this zone by root growth and burrowing animals. Zone B below has less organic matter. Here some of the chemical components may be reprecipitated, for example as clay minerals or iron oxides. The lowest zone (C) consists mainly of broken bedrock, still to be affected by the processes higher in the profile.

Many factors determine the soil type in different areas of the British Isles (Fig. 6.3), but the conditions of water flow are particularly important. High rainfall onto permeable, usually sandy, soil favours the leaching and reprecipitation that is characteristic of **podzols** (Fig. 6.4). However, a calcareous bedrock tends to neutralize acid groundwater and inhibit the leaching process, resulting in **brown soils**. In these and podzols the **water table**, below which the ground is waterlogged, is generally below the soil profile. By contrast, **gley soils** develop where some of the soil is waterlogged, giving **anaerobic** conditions that favour the formation of, for instance, iron sulphides rather than oxides. The high water table can be permanent, as on river floodplains, or a temporary result of high rainfall onto poorly permeable soil.

The nature of the bedrock strongly influences the overlying soil type, particularly its proportion of sand, silt and clay-sized particles and its resulting permeability. **Lithomorphic soils** are an extreme case, simply containing rock fragments and grains mixed with organic matter. At the other extreme, **peat soils** are dominated by plant material accumulated in waterlogged conditions.

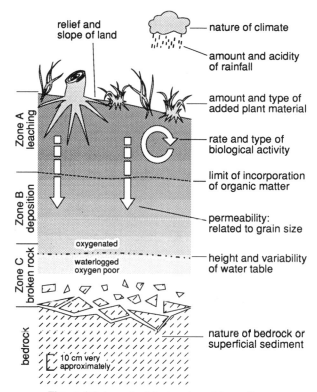

Figure 6.3 Important factors in soil formation.

Lithomorphic soils

> thin soils formed directly over solid or shattered bedrock by addition of organic material only

Brown soils

> well drained mineral soils, with water table below the soil profile; little leaching and reprecipitation of minerals

Podzols

> soils where iron and aluminium has been leached from upper levels by acid water, and redeposited in lower levels, sometimes as a cemented iron pan

Gley soils

> soils that are permanently or periodically waterlogged, giving anaerobic conditions; due to high water table (groundwater gleys) or low permeability (surface water gleys)

Peat soils

> predominantly organic soils formed by accumulation of plant material in waterlogged conditions

Human influenced soils

> soils modified or created by human activity such as dumping of refuse or manure, or land restoration

Figure 6.4 Main soil types in the British Isles.

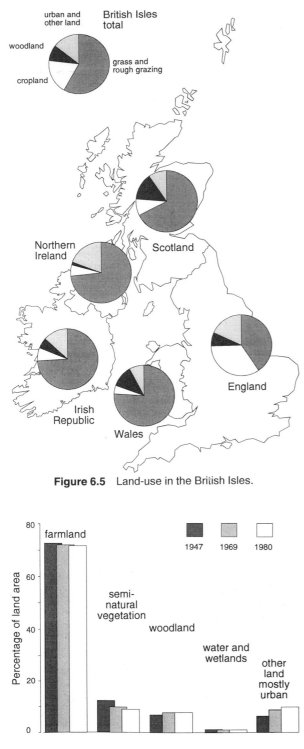

Figure 6.5 Land-use in the British Isles.

Figure 6.6 Changes in land-use in England and Wales.

6.3 Land-use in the British Isles

The strong geological and climatic contrasts between northwest and southeast Britain influence both land-scape and land-use (Chs 2 & 3, Fig. 6.5). The relatively dry lowlands that dominate England are most suitable for arable farming. Over a third of English land is used for growing crops, compared with less than 10% in Scotland, Wales or Ireland. In these northwestern regions there is a correspondingly higher proportion of grassland and rough grazing for livestock farms. The uplands of Scotland and Wales support a high proportion of Britains woodlands, many of them forestry plantations rather than natural vegetation. By contrast, England has a large amount of built-up area and land used for roads and mineral workings. In Scotland and Ireland the category of "other land" includes substantial wetlands, lakes, peat bogs and bare rock.

The balance of land-use is changing significantly in response to the pressures of development. This is most noticeable in England and Wales (Fig. 6.6). Between 1947 and 1980 the area of urban and other built-over land increased by over 50%, to about 10% of the total land area. There were corresponding decreases in the amount of farmland, wetland and, in particular, semi-natural vegetation. Woodland increased from 7% to nearly 8% of the total land area. Almost all this increase was in coniferous forest rather than in broadleaved woodland, although during the 1980s the proportion of broadleaved plantings rose significantly.

Proper planning of the future use of land requires the assessment of **land capability** (Ch. 6.1). Each projected use has its own requirements in terms of factors such as ground stability, soil quality, hydrogeology and previous land-use. The factors relevant to a particular use must be displayed on **thematic maps** derived from more general land surveys. For example, maps of clayey soils that are susceptible to shrinkage, causing structural damage to buildings, can be abstracted from more general soil surveys (Fig. 6.7). Such maps are needed on all scales, from those to guide regional development down to that relevant to a particular building site. The distribution of soil types across the British Isles (Fig 6.7a) reflects both geological and climatic factors. Most specific are the many peat soils and podsols in the wet climates of Scotland

Figure 6.7   Soil types in the British Isles and the main areas of clayey soils in Britain.

and Ireland, and the many brown soils on the limestones of England. Some agricultural impacts on soil are covered later (Ch. 15.6).

Many activities degrade land for future use (Fig. 6.2) so that, after those activities cease, it is useless without restoration. This **derelict land** accounts for about 3% of urban land in England (Fig. 6.8). This does not include the much larger land area which will be classed as derelict when its present use ends – much urban land, and the large rural area covered by motorways and roads. It is notable how much derelict land has resulted from mineral extraction, both the excavations themselves and their accompanying spoil heaps. These impacts are detailed later (Ch. 14).

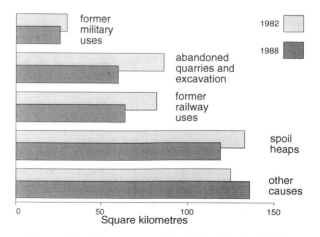

Figure 6.8   Former uses of derelict land in England.

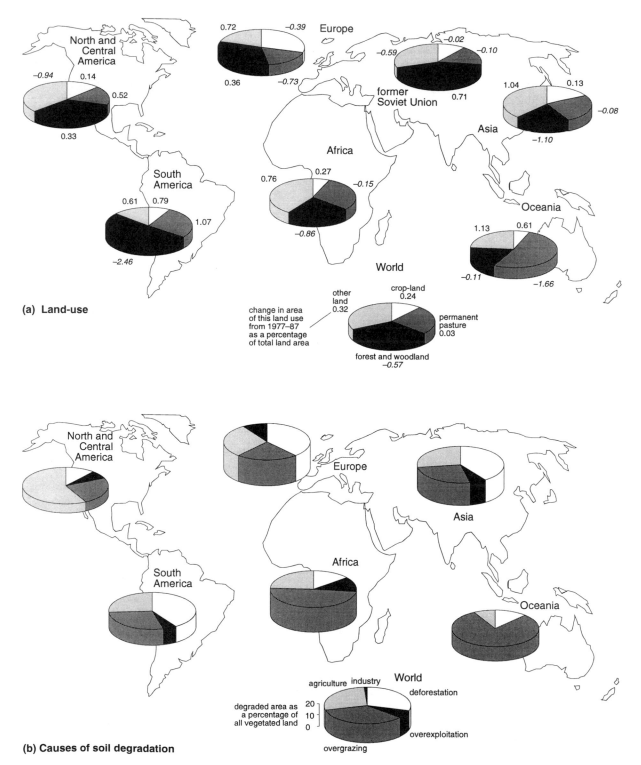

(a) Land-use

change in area of this land use from 1977–87 as a percentage of total land area

World

other land 0.32
crop-land 0.24
permanent pasture 0.03
forest and woodland −0.57

(b) Causes of soil degradation

degraded area as a percentage of all vegetated land

World

agriculture industry
deforestation
overexploitation
overgrazing

Figure 6.9   (a) Land-use and its rate of change in different regions of the world; (b) degraded land as a percentage of vegetated land in each region, together with the causes of degradation.

6.4 Global perspectives

Land-use in most of the world differs markedly from that in the British Isles (Fig. 6.9a). Nearly three-quarters of land in Britain and Ireland is farmed for livestock or crops, compared with about a third of land worldwide. Britain has correspondingly less woodland, forest and uncultivated land. These differences reflect the long history of agricultural development in Britain, with much natural forest having been cleared by the 15th century and most of the rest by the beginning of the 18th.

Similar agricultural pressures are now the major influence on changing land-use in the developing world. South America is the region most affected, with nearly 2% of the land area converted from forest to farmland over the decade to 1987. This conversion often results in degradation of the land, the loss of permanent vegetation allowing leaching of soil nutrients and major erosion of the soil cover itself.

About 17% of the Earth's vegetated areas are counted as degraded, with deforestation causing 30% of this total (Fig. 6.9b). The impact of deforestation is as high as 40% in South America, Asia and Europe. Deforested land can become wasteland, useless even for agriculture. This accounts for much of the increase in "other land" over the last decades (Fig. 6.9a).

A more important cause of land degradation even than deforestation is agricultural practice itself. Overgrazing by livestock exposes soil to wind and water erosion. It accounts for 35% of degraded land worldwide, and as much as 80% of that in Oceania. Unsustainable methods of arable farming are almost as destructive, affecting 28% of degraded land worldwide and two thirds of the North American tally. Agricultural land in the British Isles is also at risk from changes in farming methods (Ch. 15.6). Overexploitation of land for fuelwood is a further cause of soil degradation in Africa, Asia and South America (Fig. 6.9b).

A further factor in changing land-use is the growth of cities, industry and transport systems. These convert land from all other categories, but most seriously from good-quality agricultural land. Industrialization is counted as a widespread cause of land degradation only in Europe, but is a growing problem in the developing world (Ch. 15.7, 15.8).

Further reading

Avery, B.W. 1990. *Soils of the British Isles*. Wallingford, Berks.: CAB International. [A comprehensive description of soil types in both Ireland and the United Kingdom.]

Cruickshank, J. G. & D. N. Wilcock (eds) 1982. *Northern Ireland: environment and natural resources*. Belfast: Queen's University and New University of Ulster. [Chapter 8, by J. G. Cruickshank, covers soils and Chapters 6, 9, 10 and 12 other aspects of land-use.]

Department of the Environment 1992. *The UK environment*. London: HMSO. [Chapter 4 and 5 provide statistical data on soils and land-use, but with limited analysis or comment.]

Fitzpatrick, E. A. 1986. *An introduction to soil science*, 2nd edn. Harlow, Essex: Longman. [A clear introduction to the formation and classification of soils.]

Gillmor, D. A. (ed.) 1979 *Irish resources and land use*. Dublin: Institute of Public Administration. [Chapter 5, by M. J. Gardiner, covers soils and Chapters 6, 7, and 12 other aspects of land-use in the Irish Republic.]

Irish National Committee for Geography 1979. *Atlas of Ireland*. Dublin: Royal Irish Academy. [Includes a range of maps summarizing land-use in the whole of Ireland up to about 1971.]

McCall, G. J. H. & B. Marker (eds) 1989. *Earth science mapping for planning, development and conservation*. London: Graham & Trotman. [Collection of articles on the rôle of geology in land-use planning.]

World Resources Institute 1992. *World resources 1992–93*. Oxford: Oxford University Press. [Data tables and analysis of the global environment. Chapters 8 and 19 cover land-use.]

The British Geological Survey publishes thematic maps summarizing geological factors relevant to land-use. The coverage of the UK is limited, with the Midland Valley of Scotland best served.

The coverage of the British Isles by detailed soil maps is patchy, and in Northern Ireland negligible. Current availability can be checked with the appropriate body:

England and Wales: Soil Survey and Land Research Centre, Silsoe College, Silsoe MK45 4DT [1:63 360 and 1:25 000 map sheets of about 20% of the land area.]

Scotland: Soil Survey, Macauley Land Use Research Institute, Craigiebutler, Aberdeen AB9 2QJ [1:50 000 or 1:63 360 maps of about half the country.]

Ireland: Soil Survey, An Foras Taluntais, 19 Sandymount Avenue, Dublin 4 [1:126 720 maps of about a third of the country.]

Chapter Seven
WATER

7.1 The hydrological cycle

Water is one of the fundamental resources of human life. It is also abundant and renewable. Yet the lack of clean water kills at least 25 million people in the developing world each year. Water shortages constrain activity even in the most developed nations on Earth.

The total volume of water on, above and below the Earth's surface is about 1400 million km³ (Fig. 7.1). Nearly 98% of this is stored in the ocean or in seafloor rocks, suitable for human use only after energy-intensive desalination. Of the freshwater that remains, about 80% is locked in ice-sheets, also unavailable without large energy inputs. The largest reservoir of exploitable water occurs as **groundwater** in the subsurface. In the **saturated zone** water completely fills the pore spaces between rock particles. In the **unsaturated zone**, above the **water table**, water occurs as films around each particle. **Surface water** occurs in lakes and rivers. The total volume of the groundwater and surface water reservoir is over 6.3 million km³.

Most critical to the availability of water are the annual **fluxes** between the reservoirs, making up the **hydrological cycle** (Fig. 7.1). A large volume of water evaporates from the ocean and a comparable volume reprecipitates there. However some 40 000 km³ of this water is effectively transported by the atmosphere to fall over the continents. This water joins the precipitation derived from evaporation and **transpiration** by plants on land. The excess precipitation replenishes groundwater and surface-water reservoirs, before returning to the ocean, mainly as **runoff** through rivers.

It is these 40 000 km³ of water that are potentially available each year for human society to collect, use and return into the natural cycle. However, about 65% of it flows during floods when it cannot be readily trapped. Another 13% flows in uninhabitable regions, leaving about 9000 km³ as a usable resource. Averaged over the Earth, this is about 1600 m³ per person, about four times the total per capita consumption in the British Isles. Water shortages arise because neither natural rainfall nor populations are globally uniform. Rainfall varies both in time, between seasons and years, and in space: Iceland has over 20 000 times as much available water per head as Egypt.

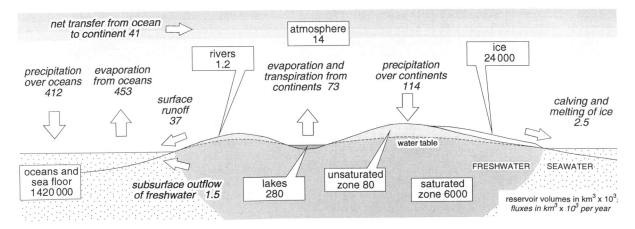

Figure 7.1 The hydrological cycle: the main water reservoirs and the fluxes between them.

7.2 Water availability and use in the British Isles

The imbalance of available water between areas and from one year to the next is evident even within the British Isles. Annual average rainfall varies from less than 500 mm in the lowlands of eastern England to over 3000 mm in the uplands of western Britain and Ireland (Fig. 7.2). This pattern results from both the southwesterly prevailing winds and the mountainous, geologically older terrains of the north and west.

This same geology means that the population density is greater both in southeast Britain and in eastern Ireland (Ch. 3). So the demand for water is greatest precisely in the areas where the rainfall is lowest (Fig. 7.3). In the north and west of the British Isles the annual demand is less than half the **effective drought rainfall** – the effective rainfall (annual rainfall less evaporation) during a one-in-fifty-years drought. However, in East Anglia demand reaches two-thirds of the effective drought rainfall and in the Thames region demand exceeds it, the deficit in dry years being made up by re-use of river water, drawdown of water reservoirs, and imports from other regions.

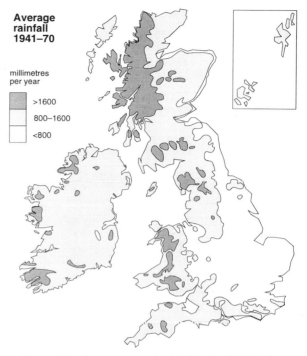

Figure 7.2 Average annual rainfall in the British Isles.

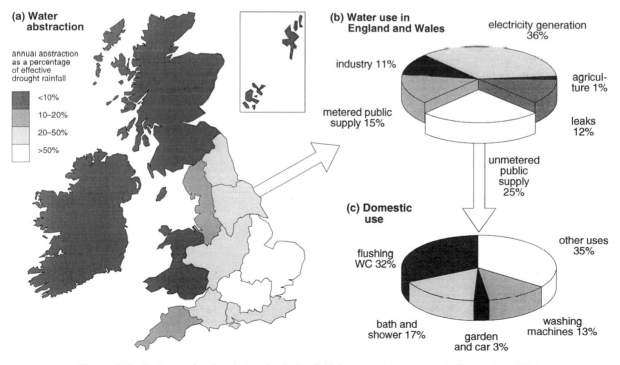

Figure 7.3 Patterns of water abstraction in the British Isles, and water use in England and Wales.

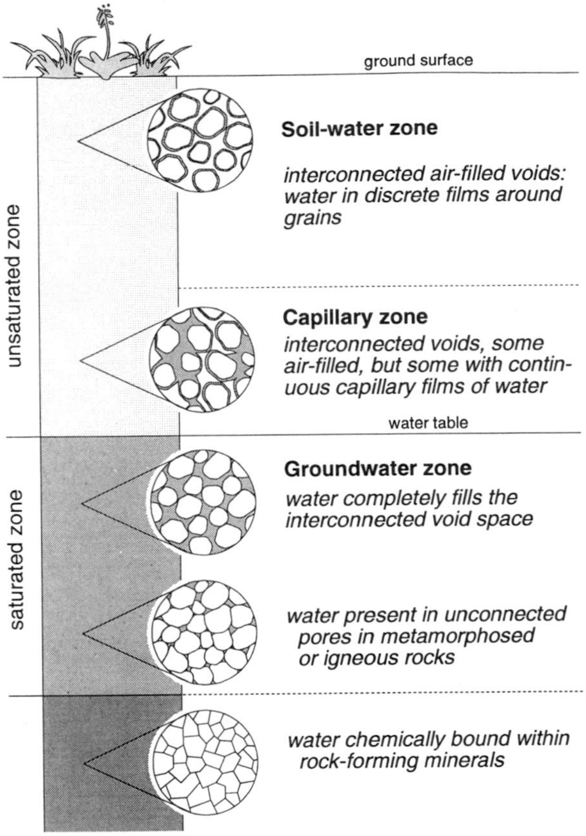

ground surface
Soil-water zone
interconnected air-filled voids: water in discrete films around grains
Capillary zone
interconnected voids, some air-filled, but some with continuous capillary films of water
water table
Groundwater zone
water completely fills the interconnected void space
water present in unconnected pores in metamorphosed or igneous rocks
water chemically bound within rock-forming minerals
unsaturated zone
saturated zone

Figure 7.4 Water in a soil/rock profile (not to scale).

7.3 Behaviour of groundwater

Groundwater is mostly supplied by rainfall. Water soaks downwards wherever the soil and rocks are **permeable**, that is with interconnected **pores** between grains. Each rainfall raises the **water table**, the level at the top of the **saturated zone** in which the pore space is filled with water (Fig. 7.4). Above the water table, in the **unsaturated zone**, air occupies part of the pore space. However, some water is stored here as films around each grain or held by surface tension in small pores. In the lower part of the unsaturated zone this **capillary action** sucks water upwards to maintain the moisture content between rainfalls, at the expense of lowering the water table.

Deep in the saturated zone, compaction narrows the pores and chemical precipitation may fill them with mineral material. Water is isolated in unconnected pores and cannot be extracted. Deeper still in the crust, rocks have negligible pore space and contain only the water that is chemically bound in their mineral structures (Fig. 7.4).

However, the availability of groundwater is not simply a matter of depth, but is strongly controlled by the variety and complexity of rocks near the Earth's surface (Fig. 7.5). **Aquifers** – permeable and porous rock units that store and transmit water – are separated by less permeable **aquicludes**. An **unconfined aquifer**,

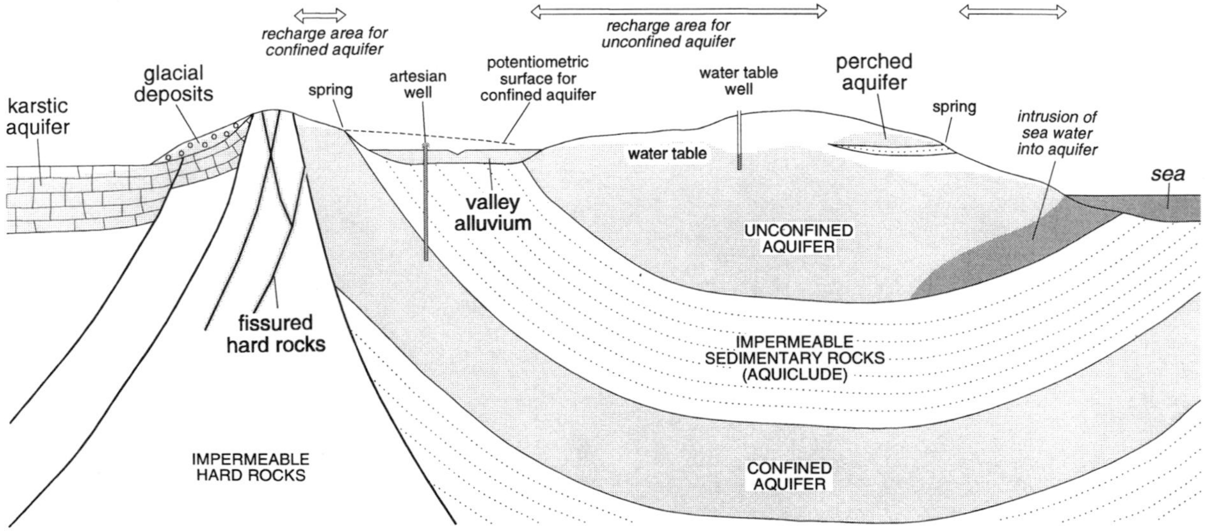

recharge area for confined aquifer
recharge area for unconfined aquifer
karstic aquifer
glacial deposits
spring
artesian well
potentiometric surface for confined aquifer
water table well
perched aquifer
spring
intrusion of sea water into aquifer
sea
water table
valley alluvium
UNCONFINED AQUIFER
fissured hard rocks
IMPERMEABLE SEDIMENTARY ROCKS (AQUICLUDE)
IMPERMEABLE HARD ROCKS
CONFINED AQUIFER

Figure 7.5 Types of aquifer (not to scale).

open to the Earth's surface, can be **recharged** directly by rain from above. A **confined aquifer**, mostly overlain by impermeable rocks, is recharged only through its unconfined portion.

An unconfined aquifer has a water table that nearly parallels the ground surface above. Water in a supply well sunk into it will rise only to the level of the water table, and must be pumped or lifted to the surface. Extraction from such a **water-table well** depresses the natural water table. In a well sunk into a confined aquifer, water will rise to a pressure level, the **potentiometric surface**, determined by the height of the recharge zone (Fig. 7.5). If this zone is on high ground, the pressure at the well site can be high enough to make water overflow at the ground surface. This is an **artesian well**. Natural overflow from a full aquifer occurs at a **spring**, where the water table intersects the ground surface.

Most aquifers are sandstones and limestones: **karstic limestones** are particularly permeable due to chemical solution (Fig. 7.5). Igneous and metamorphic rocks yield water only where they are strongly fractured. Valley alluvium and glacial deposits contain localized but common near-surface aquifers. Water can also be trapped in **perched aquifers** above thin aquicludes in the unsaturated zone.

All aquifers are at risk of pollution from human activity (Ch. 16). The main natural pollutant is salt water, which intrudes downwards as a plume in any aquifer that is exposed on the sea floor.

7.4 Management of surface water

Geologists are most concerned in the exploitation of groundwater. However, surface water provides the major supplies in most of the British Isles. Small communities can be supplied directly from a spring or stream, but reservoirs allow seasonal and inter-annual variations in flow and rainfall to be evened out more reliably (Fig. 7.6). Towns can be supplied by pipeline or aqueduct directly from a reservoir, or through a river whose flow is maintained at a steady level by the input of surplus water. This water might be from a reservoir upstream on the same river system, one in an adjacent drainage basin, or from an aquifer. Conversely, reservoir water can be used to recharge aquifers where rainfall is insufficient to balance extraction.

Reservoirs are usually constructed in the deep valleys of upland areas. Lowland reservoirs have a greater surface area and evaporation rate for a given volume of stored water. A reservoir can also be created at sea level behind an **estuarial barrage**. **Desalination** of sea water is a last resort for most countries due to the high costs of energy involved.

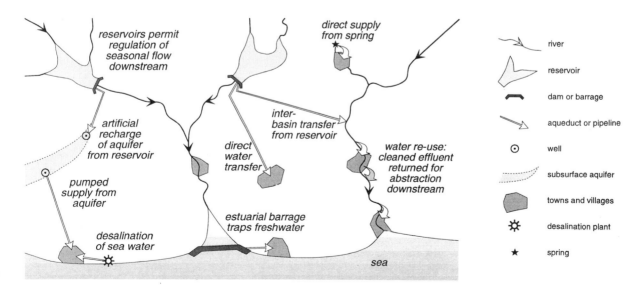

Figure 7.6 The management of surface water.

7.5 Groundwater in the British Isles

Groundwater provides about 18% of the fresh water used in the British Isles. However, its relative importance varies strongly across the region, from as low as 1% in Wales up to 50% in southern England (Fig. 7.7a). High reliance on groundwater is restricted to the area southeast of the Tees–Exe Line, where the dominantly post-Carboniferous sequences contain several major aquifers (Fig. 7.7b).

The *Upper Cretaceous Chalk* is the most important of these aquifers. It is a very fine-grained limestone with up to 50% porosity, but with a low permeability between the small grains. Fractures, mainly joints, provide the main channels for water flow, increasing the effective permeability by a hundred times or more.

The Chalk underlies or crops out over much of southern and eastern England. In the south it is affected by gentle east-trending folds, and has been uplifted and eroded in the Weald Anticline. The syncline of the London Basin provides the main source of groundwater for London. Recharge on both the southern and northern limbs of this structure originally maintained an artesian flow. However, over-abstraction has severely lowered the water level in the Chalk aquifer (Ch. 14.2). London also relies heavily on surface water from the Rivers Thames and Lea.

A more typical scheme of water extraction from the Chalk is exemplified by the unconfined aquifer in the Cambridge area (Fig. 7.8). A Cretaceous sequence of Lower Greensand, Gault Clay and Chalk overlies the Jurassic Kimmeridge and Ampthill Clays. The strata

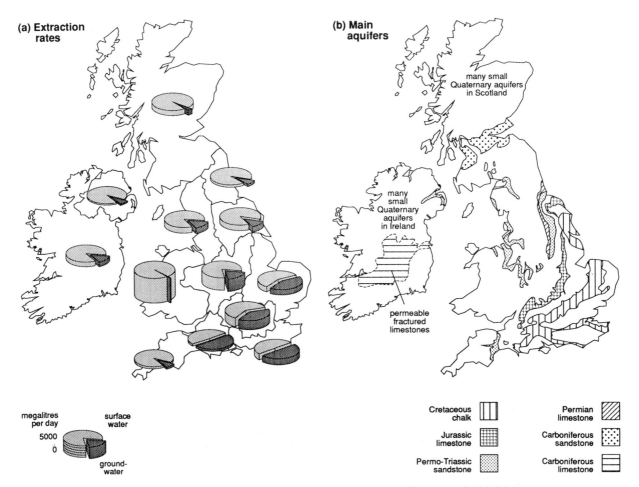

Figure 7.7 Groundwater extraction rates and the main aquifers in the British Isles.

dip gently southeastwards, so that low ground on the clays in the northwest is bordered by hills on the Chalk. The water table in the Chalk rises to the southeast in sympathy with the ground surface. The lowest part of the Chalk is clay-rich, less fractured and therefore less permeable than higher parts. The first public water supply for Cambridge tapped natural springs that issued at this permeability transition within the Chalk. Now a series of boreholes penetrate the permeable chalk, down to about 80 m below ground level, where the fracture permeability decreases. Water from these boreholes is pumped northwestward to Cambridge, then northward to an area lacking groundwater and with the lowest rainfall in Britain. Some extracted water is used to augment the flow in rivers that were originally fed by springs now weakened by a lowered water table. Part of this augmented river flow is abstracted downstream and piped southwards to the Essex rivers and east London.

Permo-Triassic sandstones form the second most important aquifers in the British Isles. Although their porosity is lower than that in the Chalk, the sand particles permit a good intergrain flow. However, fractures are again important in enhancing permeability locally. The Permo-Triassic aquifers occur both in east-dipping sequences, such as in northeast and southwest England, and around structural basins, such as in the English Midlands, southern Scotland and Northern Ireland (Fig. 7.7b). Associated evaporites make the water too salty to use in places.

Other *pre-Quaternary aquifers* are only locally important, when compared with the 75% of ground water derived from the Chalk and the Permo-Triassic sandstones. The more significant are the Magnesian limestone, at the base of the Permian sequence in northeast England, the thicker Jurassic limestones, and some Cretaceous sandstones (Fig. 7.7b). Carboniferous sandstones are important in Scotland, and Lower Carboniferous limestones in Ireland, particularly where they are strongly fractured in the southern part of their outcrop.

Quaternary aquifers in glacial and river gravels provide local small-volume supplies throughout the British Isles, but are particularly valuable in Ireland and Scotland. Quaternary deposits are also crucial in allowing or restricting recharge to deeper aquifers.

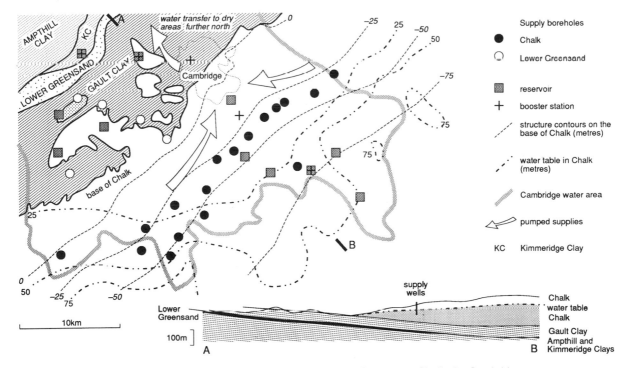

Figure 7.8 Example of groundwater extraction from the Cretaceous Chalk: the Cambridge area.

7.6 Water futures in the UK

The British Isles are not short of water. The UK utilizes only 25% of its groundwater and less than 10% of its surface water. Estimates for Ireland are 2.5% and 2% respectively. However, the regional imbalance of available water (Fig. 7.3) means that future supply strategies must be carefully planned, particularly for the drier, more populated areas of eastern Ireland and southeastern England.

Demand for water in England and Wales is growing only slowly, by about 4% over the decade 1980–1990 (Fig. 7.9). A steady increase in piped mains supply, mostly for domestic purposes, was balanced by a decrease in industrial use. This decrease, due to the decline of heavy industry, was unforeseen by the water-supply authorities. Reservoirs were built to satisfy anticipated industrial growth, so that most northern and western regions had a large surplus of supply capacity by the end of the decade (Fig. 7.10). Predicted increases in demand over the next thirty years leave only the southeast short of water. There is a vigorous debate about whether to meet this shortfall by constructing more reservoirs, by pumping surplus water from the northwest, or by stemming the present 25% leakage of mains water from broken pipes.

Some surplus of supply capacity over average demand is maintained to allow for occasional years of drought. A new factor in this assessment is the risk of climate change induced by global warming (Ch. 17). The current predictions are that the beneficial effect of wetter winters in a warmer British Isles climate would be offset by greater summer evaporation and more climatic variability.

7.7 Global perspectives

The uneven pattern of water in the British Isles is a microcosm of a more serious global inequality in water resources. The availability of water per person varies over four orders of magnitude (Fig. 7.11) from the negligible supplies in parts of the Middle East to the large surpluses of the tropics and some high-latitude countries. An inequality occurs through time as well as space, with drier areas being dependent on seasonal rainfall, which is more difficult to manage for year-round use. Drier areas are also more prone to intermittent drought. Water shortages encourage re-use of extracted water, with attendant risks of pollution (Chs 13, 15, 16).

The balance of water use in the British Isles (Fig. 7.3a,b) is typical of the developed world only. About 70% of the world's abstracted fresh water is used for agriculture, 20% for industry and less than 10% for domestic purposes. Agricultural use reaches 90% in many African countries. Failure of supplies in such areas is more likely than in the developed world to produce food shortages and severe human suffering.

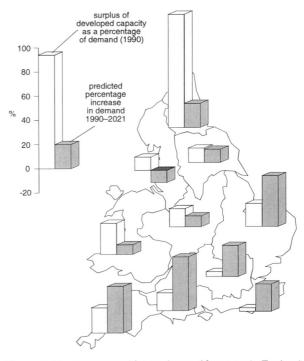

Figure 7.10 Present and future demand for water in England and Wales.

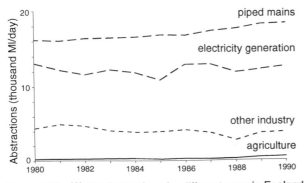

Figure 7.9 Water abstractions for different uses in England and Wales during 1980–90.

Water-supply problems in developing nations are exacerbated by lack of money to build the infrastructure of water catchment, purification and transmission. As a result, wealthy countries tend to use more water per person even where its availability is lower (Fig. 7.12). For instance, the Middle Eastern oil states such as Kuwait and Saudi Arabia can afford to extract adequate fresh water by energy-intensive desalination processes where natural resources are scarce. By contrast, Ethiopia, which has as much available water per person as the United Kingdom, extracts less than a tenth as much per head. Water shortage lowers the agricultural and economic prosperity of such countries, creating a cycle of water-supply difficulties. Breaking this cycle is costly, technically demanding and often politically sensitive.

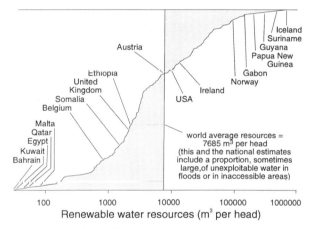

Figure 7.11 World water availability by country.

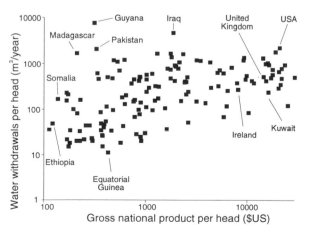

Figure 7.12 Water use plotted against wealth.

Further reading

Brassington, R. 1988. *Field hydrogeology*. Milton Keynes, Bucks.: Open University Press. [A guide linking practical techniques to hydrogeological and hydrological theory.]

Clarke, R. 1991. *Water: the international crisis*. London: Earthscan. [An overview of global problems of water supply.]

Cruickshank, J. G. & D. N. Wilcock (eds) 1982. *Northern Ireland: environment and natural resources*. Belfast: Queen's University and New University of Ulster. [Chapter 3, by R. Common, covers water supply and demand, and Chapter 7, by J. C. Roberts, includes a section on aquifers.]

Department of the Environment 1992. *The UK environment*. London: HMSO. [Chapter 6 provides statistical data on water resources and abstractions, but with little analysis or comment.]

Downing, R. A. & W. B. Wilkinson 1991. *Applied groundwater hydrology*. Oxford: Oxford University Press. [An excellent collection of papers covering problems of both supply and potential pollution.]

Gillmor, D. A. (ed.) 1979 *Irish resources and land use*. Dublin: Institute of Public Administration. [Chapter 3, by D. P. Drew, reviews water resources in the Irish Republic.]

Kirby, C. 1984. *Water in Great Britain*. Harmondsworth, Middx: Penguin. [A useful, readable review.]

Pearce, F. 1982. *Watershed: the water crisis in Britain*. London: Junction. [A critical analysis of water-supply problems, mainly in England and Wales.]

Pearce, F. 1992. *The dammed: rivers, dams and the coming world water crisis*. London: Bodley Head. [A perceptive review of conflicting solutions to water supply problems throughout the world.]

Price, M. 1985. *Introducing groundwater*. London: Allen & Unwin. [A readable but authoritative introduction to hydrogeological principles.]

Robins, N. S. 1990. *Hydrogeology of Scotland*. Keyworth, Nottingham: British Geological Survey. [Results of a reconnaissance of groundwater potential in Scotland; detailed but not too technical.]

Smith, K. 1972. *Water in Britain*. London: Macmillan. [An introduction to the hydrology and geography of Britain's water supply.]

World Resources Institute 1992. *World Resources 1992-93*. Oxford: Oxford University Press. [Data tables and analysis of the global environment. Chapters 11 and 22 cover fresh water.]

Chapter Eight

CONSTRUCTION MATERIALS

8.1 Use of construction materials

Geological materials have been used for building for much of human history. Strong rocks form **dimension stone** for walls (Fig. 8.1). Thin-bedded **flags** or cleaved **slates** are suitable for roofs. Clay can be shaped, then sun-baked or kilned to make **bricks** and **tiles**. Clay can be mixed with limestone, then fired to make **cement**. Cement, sand and water form **mortar** for bonding stone and brick. Gypsum is heated to produce **plaster**. Only timber matches rocks in importance as a traditional building material.

Industrialization brought increased use of geological materials (Fig. 8.1). Pure silica sands are melted and cast into **glass** for windows. Unconsolidated sand and gravel, known as **aggregate,** is mixed with cement and water to produce **concrete**. Better tecchniques to strengthen this material and to pour it into place have revolutionized the architecture and construction of buildings, and of engineered structures such as bridges, dams and tunnels. Concrete is also cast into building blocks, paving slabs, kerbstones, pipes, and many other components of the built environment.

Dimension stone is now used for facing buildings as well as for for load-bearing construction. Large blocks are used as **armourstone** facing to coastal defences or river banks. But most hard rock is crushed to form **aggregate**. Some of this goes for concrete but more is used as **roadstone** either loose in the lower layers or, mixed with bitumen, as a hard-wearing surface layer. Aggregates of all origins are extensively used as **fill**, for example in the foundations of structures and in the cores of dams.

Construction materials are relatively common substances. This distinguishes them from **industrial** and **metallic minerals** (Ch. 9) which occur in local concentrations formed by less widespread geological processes. However, even common rocks rarely occur where they are to be used. Being bulky and dense, they are costly to transport, so proper assessment is needed to find deposits that are both economically and geologically suitable.

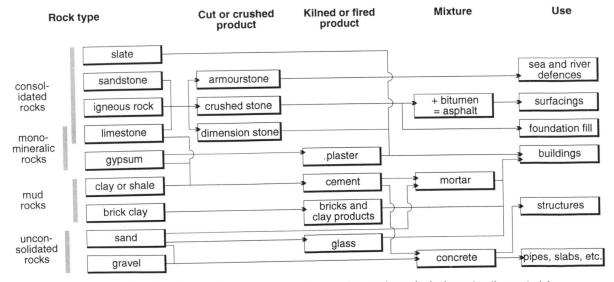

Figure 8.1 Flow chart showing the processing and uses of the main geological construction materials.

8.2 Properties of construction materials

Cement (strictly, **Portland cement**) is a mixture of calcium carbonate ($CaCO_3$), silica (SiO_2) and alumina (Al_2O_3), kilned at high temperature to produce a range of calcium silicates and aluminates. Limestone provides the carbonate but usually requires the addition of about 20% clay, shale, or **pulverized fly ash** from coal-fired power stations, to supply enough alumina and silica (Fig. 8.2). Some impure limestones may have the right chemical mixture to be used directly, but these **cement rocks** are localized and uncommon.

The limestone and shale are crushed, ground to a powder, and mixed either dry or in a water-based slurry (Fig. 8.2). This mixture is fed through a rotary kiln fuelled by powdered coal or gas. Water and carbon dioxide are driven off at low temperatures before reactions at up to 1450°C form a silicate and aluminate **clinker**. The cooled clinker is mixed with about 5% of gypsum, which moderates the setting time of the final product, and then powdered for use.

Plaster is a type of calcium sulphate ($CaSO_4 . \frac{1}{2}H_2O$) formed by dehydrating **gypsum** ($CaSO_4 . 2H_2O$) at about 107°C. Gypsum occurs naturally in evaporite deposits (Ch. 9) and is increasingly generated as a byproduct of **flue gas desulphurization** (FGD) of gases from power stations (Ch. 17).

Bricks and other *clay products* are produced by moulding suitable clay-bearing rocks to the required shape, then drying and firing them to form the high-temperature silicates, oxides and glass which give clay products their rigidity and durability. The original mineralogy of the clay, shale or slate is crucial to the success of the process (Fig. 8.3). The proportion of clay minerals and water controls the plasticity of the rock during moulding. The clay minerals provide the alumina for the high temperature reactions, but quartz is important in providing strength and resistance to weathering. The most valuable components of a good brick clay are its contained organic particles. These provide internal fuel to each brick during firing, giving a more even temperature and substantially reducing the energy input and cost of the process. The distinctive colours of bricks (Ch. 3) are affected by the proportion of iron-bearing minerals and by whether the kiln atmosphere is controlled to be oxidizing or reducing.

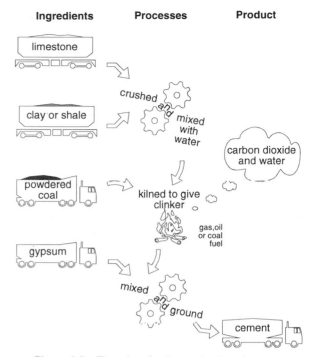

Figure 8.2 Flow chart for the production of cement.

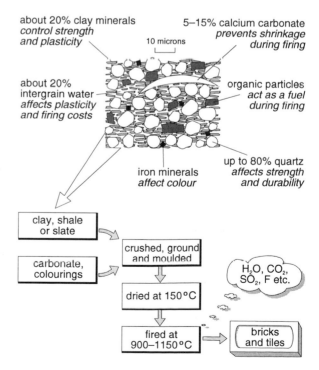

Figure 8.3 Optimum properties of a brick clay.

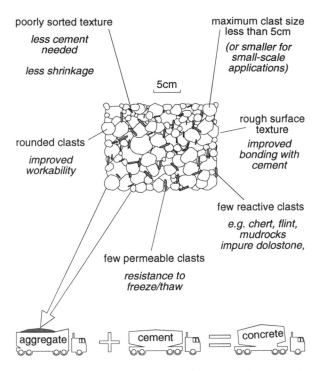

Figure 8.4 Optimum properties of aggregate for concrete.

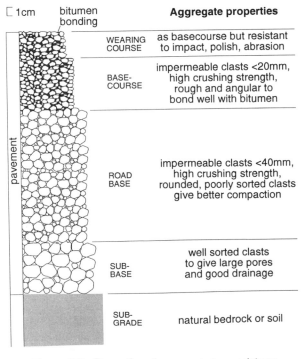

Figure 8.5 Properties of aggregate for roadstone.

Aggregate is particulate construction material of sand size and above. Aggregate can be uncemented sand and gravel or consolidated rock that has been crushed to size. Industrial wastes such as crushed concrete, colliery or slate waste, and furnace slag can be substituted for some applications. However, each use requires aggregate with specific properties.

Concrete is a mixture of coarse aggregate with cement and water. The cement minerals react with water to form interlocking crystals of hydrated silicates and aluminates, binding the aggregate into a solid material. Good aggregate has rounded particles, to slide over each other and make the concrete more workable as it is poured, and a rough surface texture, to improve the bonding with the cement crystals (Fig. 8.4). A poorly sorted aggregate needs less cement, because the smaller grains fill the pores between the large ones. However, permeable particles are unwanted because they make the concrete prone to freeze/thaw weathering. Also undesirable are clasts of chert, flint, impure dolostones or mudrocks, which react with the cement and cause expansion and cracking. The optimum properties for concrete aggregate are found in glacial, fluvial and marine deposits of sand and gravel. Crushed rock, being angular, is less suitable, but is used in areas where gravels are scarce.

Mortar, *cement screed* and *rendering* are mixtures of cement with sand, used respectively for binding bricks, and coating floors and walls. Rounded sand grains improve the workability of the mixture, favouring so-called **soft sands** over the **sharp sands** which are acceptable in concrete or road aggregate.

Road aggregate makes up the various layers of a road **pavement** (Fig. 8.5). The **sub-base** is a coarse drainage layer, with large pores to prevent capillary rise of water. The **roadbase** bears and spreads the traffic load, and ideally needs strong, rounded, poorly sorted particles. These may or may not be weakly bonded with bitumen. The upper two layers are always bitumen-bonded, so that a rough surface texture and angular shape to the particles are important as well as strength. Aggregate for the **base-course** can be to a less stringent specification than for the **wearing course**, where high resistance to abrasion and polishing are essential. Crushed igneous rocks and coarse or indurated sandstones make the best wearing-course aggregates.

Rock fill is aggregate used for purposes where strength and size specifications are not as critical as for concrete or road surfacing. These uses include embankments and the cores to earth dams.

Glass is made from silica sand with a very strict specification (Fig. 8.6). Glass is mostly SiO_2 with added Na_2O (12–16%), CaO (5–11%), MgO (1–3%) and Al_2O_3 (1–3%). The most troublesome impurity is iron, which colours glass green or even brown, above an Fe_2O_3 content of about 0.2%. Most sands have grains coated with iron minerals or clay films, and naturally occurring glass sands were either deposited in unusual iron-free environments or have been leached clean by acid groundwater. Even these sands must be sieved to achieve a narrow range of grain size, essential for even melting during production.

Dimension stone is rock which can be cut or dressed into regularly shaped pieces. It was formerly used for building load-bearing walls, particularly limestones which are easy to work. However, its high cost compared with brick and concrete have made it more popular now as a facing material with a mostly decorative function. A high compressive strength is needed for building (Fig. 8.7). Many crystalline rocks and well cemented sedimentary rocks are suitable, provided that they contain few discontinuities such as bedding, joints or cleavage. These weaknesses limit the size and shape of cut blocks and increase the rate of weathering of the stone. The durability of the stone is also reduced if it has a porous texture or permeable grains which promote freeze/thaw weathering, or soluble mineral grains which suffer chemical attack in an acidic atmosphere.

Armourstone is durable rock with widely spaced bedding and joints that allow it to be extracted in large blocks. It is used for facing dams, river embankments, and breakwaters, where it needs to withstand high-energy conditions.

Flagstone and *slate* are building stones which can be split into thin sheets for flooring or roofing. Flags split along a primary sedimentary lamination, due to a strong parallel arrangement of grains or to fine alternations of laminae of differing grain size (Fig. 8.8). Slates split not on bedding but along a secondary tectonic cleavage, formed by rotation of existing grains or growth of new grains during deformation and low-grade metamorphism.

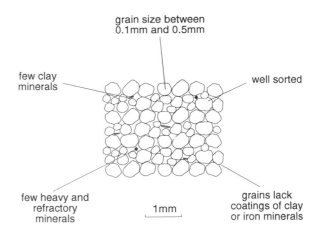

Figure 8.6 Optimum properties of silicon sand for glass.

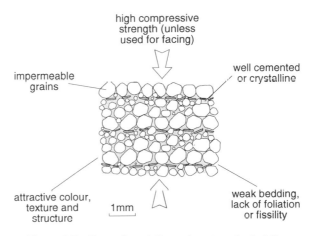

Figure 8.7 Properties of dimension stone for building.

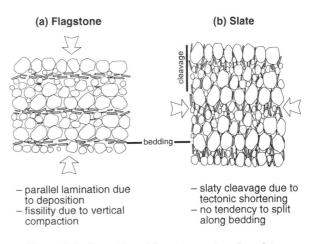

Figure 8.8 Properties of flagstone and roofing slate.

55

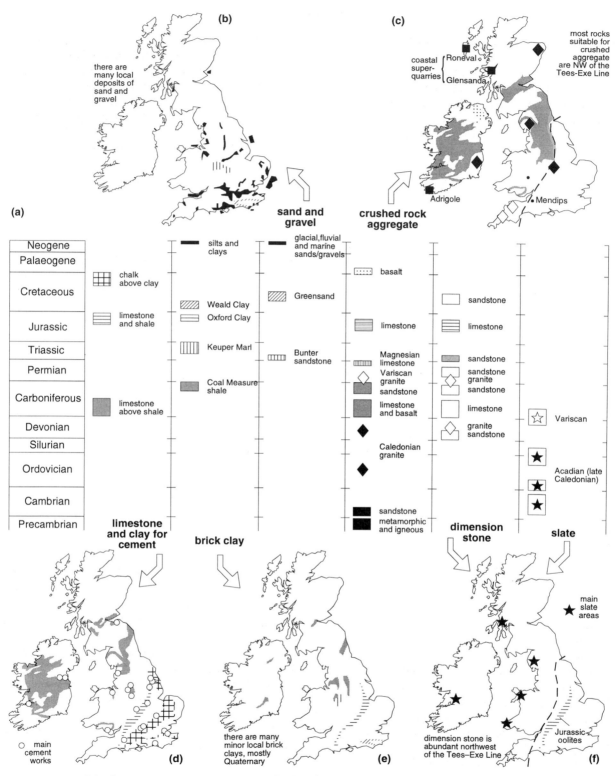

Figure 8.9 Stratigraphic and geographic distribution of the main construction materials in the British Isles.

56

8.3 Distribution in the British Isles

There is a strong stratigraphic control on the occurrence of geological construction materials in the British Isles (Fig. 8.9a). The older rocks tend to be more indurated, metamorphosed and deformed, and to contain more igneous lithologies. Therefore, Carboniferous and older units yield most of the hard rocks for crushed aggregate, whereas Carboniferous and younger units provide the best cement materials, brick clays, sand and gravel. Dimension stone is more evenly distributed through geological time.

This stratigraphic constraint leads to a marked geographic pattern to the availability of raw materials. Carboniferous and older rocks dominate Britain to the northwest of the Tees–Exe Line (Fig. 8.9c), whereas younger rocks occur mainly to the southeast. The effects of this divide on **vernacular architecture** have already been discussed (Ch. 3). Regional differences in building styles have been progessively suppressed by the easier transport of building materials, first by rail and later by road. However, the uneven occurrence of materials continues to have important economic and land-use consequences.

Cement manufacture uses limestone from three main stratigraphic levels (Fig. 8.9a). Upper Cretaceous Chalk sources about 55% of British Isles production, Lower Carboniferous limestone another 30% and Jurassic limestone most of the rest. Interbedded shales in the Jurassic provide the clay content. However, most cement works are located along the contact of the Chalk with underlying Cretaceous or Jurassic clays, or where Upper Carboniferous shales overlie the Lower Carboniferous limestones (Fig. 8.9d). Fuel has to be imported from coalfields to all these sites. The gypsum additive is mostly brought from Permo-Triassic evaporites in Nottinghamshire, Cumbria or Kingscourt (Ireland), or from the late Jurassic of the Weald (Fig. 8.9d).

Brick clays from Quaternary deposits were the first to be exploited; boulder clays, alluvial and lacustrine clays and the wind-blown silts of southeast England. However, most of these deposits are of limited extent. Improved transport for the finished product favoured increasing exploitation of older, more extensive deposits supplying larger brickworks. The Cretaceous Weald Clay, the Triassic Keuper Marl and the Upper Carboniferous shales; all these are important sources of modern brick-clays (Fig. 8.9a,e). But nearly half the bricks in Britain are produced from one unit alone, the Jurassic Oxford Clay. This has a high hydrocarbon content, which improves the plasticity of the clay during moulding and, more importantly, reduces the cost of firing by about 75%.

Sand and gravel aggregates come most readily from unconsolidated Quaternary sources (Fig. 8.9a). Some rivers in the British Isles carry and deposit economic quantities of gravel. However, many of the large accumulations date from during or immediately after the Quaternary glaciations, when the sediment load of these rivers was far higher. Former floodplains are now preserved as **terraces** above present river level, or as offshore gravel patches, flooded by sea-level rise. Some of these **fluvioglacial** gravels form extensive sheets isolated from modern river systems. The ice-deposited **till** or **boulder clay** may itself be used as aggregate, but requires washing to remove the high proportion of clay. Some ancient sands and gravels have remained weakly cemented enough to be dug as unconsolidated aggregate. These include the Cretaceous Greensand of southeast England and the Triassic Bunter sandstones and pebble beds of the English Midlands (Fig. 8.9a,b).

Crushed rock aggregate is produced mostly from rocks of Carboniferous and earlier age (Fig. 8.9a,c). The main exceptions are Tertiary igneous rocks, particularly the Northern Ireland basalts, the Permian Magnesian Limestone of northeast England and the Variscan granites of the southwest. Jurassic limestones are extensively quarried, but are too soft for demanding applications. Limestones form about 65% of all crushed rock aggregate in the British Isles, and the majority of this is from the widespread Carboniferous limestones. These deposits in the Mendips and Peak District, together with Caledonian igneous rocks in Leicestershire and Precambrian rocks in the Welsh Borderland, are important sources close to the high-demand areas of southeast England. The vast reserves of suitable Precambrian rocks in Scotland and northwest Ireland have so far proved too remote for extensive exploitation.

The distributions of *dimension stone* and *slate* have already been described (Ch. 3), but are summarized in Figures 8.9a,f.

8.4 Production in the British Isles

In many towns and villages in the British Isles, stone, slate, brick and tile still seem the dominant building materials. In fact, together, these account for less than 4% of the geological materials used for construction, with dimension stone and slate a mere 0.25%. These traditional materials are conspicuous in modern buildings only because they face or roof structures mainly built of concrete. Concrete absorbs about 28% of raw rock materials in Britain as aggregate and another 7% as cement (Fig. 8.10). Roadstone aggregate uses 27% and other constructional aggregates about 30% of this total.

Raw materials other than aggregate are produced in small quantities by comparison (Fig. 8.11). However, some specialized materials are economically more important than their tonnage suggests. Silica sand, for instance, is worth over twice as much per tonne as common aggregate. Materials for making cement and bricks are worth rather less than aggregate. These relative costs are those at the quarry gate, before the considerable cost of transport. The cost of raw aggregate effectively doubles at only 10 km from the quarry, so that there are strong incentives to find materials close to the site where they are to be used.

The growing pressure on aggregate resources in Britain is graphically shown by the rise in its production this century (Fig. 8.10). Demand in the UK reached over 5 tonnes per head in the late 1980s, much of it in southern and eastern Britain. Here there are few sources of crushed rock (Fig. 8.9). Many sand and gravel reserves are already built over or underlie valuable recreational and agricultural land. Impacts of aggregate extraction will be examined later (Ch. 14) and some issues of land-use planning have already been discussed (Ch. 6). One controversial strategy for supplying southeastern Britain is to exploit the abundant hard-rock resources of Scotland and Western Ireland in "coastal superquarries" loading aggregate directly onto ships for relatively cost-effective transport (Fig. 8.9c).

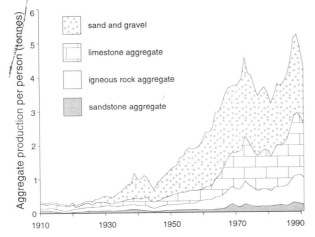

Figure 8.10 UK production of aggregate 1910–90.

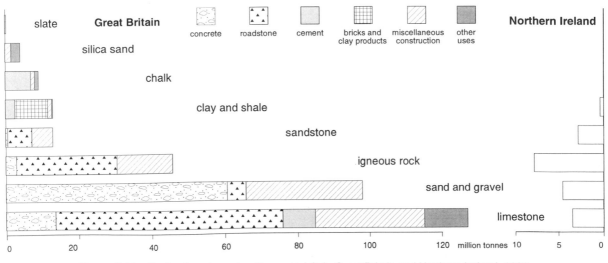

Figure 8.11 Production of construction materials in Great Britain and Northern Ireland, 1989.

8.5 Global perspectives

Construction materials are geologically common, and there are no fears of global shortages as there are for metallic minerals or fossil fuels. Of greater concern are local imbalances of supply and demand, such as those in the British Isles. Rapidly developing nations face a growth in demand comparable with that in the UK since 1945. Anticipating this growth is essential to balancing needs against environmental impacts.

Analysis of global production and use of construction materials is hampered by the lack of reliable data. Even in the British Isles, rock that is dug and used locally, for instance from farm quarries, never appears in national statistics. This pattern is normal in most of the developing world.

Cement is the one relevant commodity with a global database and is an index to other rock materials. Each tonne of cement needs an equivalent amount of limestone and a further 6 tonnes of aggregate to make it into concrete. The production of cement per head is not, as might be expected, proportional to the per capita wealth of a nation (Fig. 8.12). Few nations produce more than 0.8 tonne of cement per head. This "ceiling" has the helpful implication that rapid growth in demand for construction materials is a discrete phase in national economic development. This may be small comfort considering the vast tonnages used by developed nations.

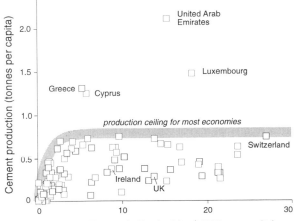

Figure 8.12 The correlation between cement production and wealth, 1988.

Further reading

Advisory Committee on Aggregates (the Verney Committee) 1976. *Aggregates: the way ahead.* London: HMSO. [A strategic survey of the demands for construction materials in the British Isles, and of solutions such as coastal superquarries.]

British Geological Survey 1993. *United Kingdom minerals yearbook 1992.* Keyworth, Notts.: British Geological Survey. [Annual statistical data and a brief review of UK production and trade in minerals, including construction materials.]

Hammond, M. 1990. *Bricks and brickmaking,* 2nd edn. Aylesbury: Shire. [A concise introduction to bricks.]

Holland, C. H. (ed.) 1981. *A geology of Ireland.* Edinburgh: Scottish Academic Press. [Chapter 18, by G. D. Sevastopulo, reviews economic geology including construction materials.]

Irish National Committee for Geography 1979. *Atlas of Ireland.* Dublin: Royal Irish Academy. [An excellent compilation, including distribution maps of mines and quarries.]

Lumsden, G. I. (ed.) 1992. *Geology and the environment in western Europe.* Oxford: Oxford University Press. [Includes a discussion of construction materials in a European context.]

Open University 1974. *The Earth's physical resources. Block 4: Constructional and other bulk materials.* Milton Keynes, Bucks.: Open University Press. [Excellent summary of geological principles, industrial processes, and engineering uses.]

Prentice, J. E. 1990. *Geology of construction materials.* London: Chapman & Hall. [Links geological processes and engineering requirements in a European perspective.]

Shadman, A. 1989. *Stone: an introduction.* New York: Bootstrap Press. [A lively mix of geology and practical advice on using dimension stone.]

Smith, M. R. & L. Collis 1993. *Aggregates: sand, gravel and crushed rock aggregates for construction purposes.* London: Geological Society. [An informative manual on the location, sampling and testing of aggregates.]

Stanier, P. 1985. *Quarries and quarrying.* Aylesbury: Shire. [Introduction to methods of working dimension stone and crushed rock.]

Stone Industries 1987. *Natural stone directory,* 7th edn. Maidenhead, Berks.: Ealing Publications. [An informative compilation of types of dimension stone and their active sources in Britain and Ireland.]

Williams, M. 1991. *The slate industry.* Aylesbury: Shire. [Introduction to extracting and processing slate.]

Chapter Nine

INDUSTRIAL AND METALLIC MINERALS

9.1 Nature and use

This chapter considers geological materials valued for their content of a specific mineral or metal. These contrast with construction materials (Ch. 8), which utilize the bulk properties of rocks, many of which contain a mixture of several minerals. However, this distinction is blurred. Most monomineralic rocks such as limestone, dolomite, chalk, gypsum and silica sand have both constructional uses (Fig. 8.1) and other industrial uses (Fig. 9.1c).

The scarcity of the rock is an important distinction. Construction materials are common rocks, whereas mineral deposits are concentrations formed by special associations of geological processes (Ch. 8.2). Mineral deposits require a correspondingly greater level of geological exploration and appraisal.

Industrial minerals are valued for their own properties rather than for any constituent metals (Fig. 9.1). Although less glamorous than metallic minerals, they are equally important in the national economies of the British Isles. Moreover most industrial minerals, except phosphate and barytes, are produced domestically whereas most metals are imported. So the value of industrial minerals produced in the UK is about a hundred times that of produced metallic minerals. China clay is the most valuable geological export from the United Kingdom after petroleum.

(a) UK consumption	(b) Main mineral	(c) Main uses
limestone	calcite $CaCO_3$	filler (animal feed, polymers, paint, paper, pharaceuticals), agriculture, flux (iron, steel, glassmaking), chemicals
salt	halite $NaCl$	chemicals
silica sand	quartz SiO_2	glass, foundry moulds, abrasives
dolomite	dolomite $CaMg(CO_3)_2$	agriculture, iron and steel, glassmaking, abrasives
gypsum	gypsum $CaSO_4.2H_2O$	filler for textiles
chalk	calcite $CaCO_3$	filler (as limestone), agriculture, iron and steel, abrasives, chemicals
fire clay	kaolinite $Al_4Si_4O_{10}(OH)_8$	refractory bricks, pipes, tiles
potash	sylvite KCl, langbeinite $K_2SO_4.2MgSO_4$	fertilizer, chemicals
china clay	kaolinite $Al_4Si_4O_{10}(OH)_8$	filler and coating for paper, porcelain
phosphate	apatite $Ca(PO_4)_3(F,Cl,OH)$	fertilizer
barytes	barite $BaSO_4$	filler for paper, drilling mud
ball clay	kaolinite $Al_4Si_4O_{10}(OH)_8$	ceramics
fuller's earth	montmorillonite $(Al,Mg)_4Si_8O_{20}(OH)_4.nH_2O$	filler for paints, drilling mud
fluorspar	fluorite CaF_2	flux (steelmaking), aluminium smelting

Scale for (a): tonnes per year, 10^5, 10^6, 10^7

Figure 9.1 Some important industrial minerals: UK consumption (1991), mineralogy and main uses.

Most common industrial minerals are monomineralic sedimentary rocks (Fig. 9.1a): carbonates, clays or **evaporites** (halite, gypsum, potash; Ch. 9.2). Barytes and fluorspar occur in crystalline **veins**, often as subsidiary or **gangue** minerals to metallic minerals.

Industrial minerals are less conspicuous in their end use than are metals. Their uses as inert **fillers** in paper, textiles, paint and plastics are prime examples (Fig. 9.1c). Minerals additional to those listed are important as hard abrasives, pigments and gemstones.

Metallic minerals are those exploited mainly for their contained metals (Fig. 9.2b,c). These metals must be extracted from the mined rock, the **ore**, by processes such as flotation, magnetic separation, leaching and smelting. The common ore minerals are either oxides, sulphides or carbonates. Gold, silver and platinum also occur as the native metal.

The abundance of economically important metals varies over more than seven orders of magnitude (Fig. 9.2e). Naturally, society uses more of the common metals than it does of the scarce ones (Fig. 9.2a), but the correlation is inexact. Zinc, copper and lead stand out as more heavily consumed than their crustal abundance might suggest. This has implications for the projected lifetime of their reserves (Ch. 9.5).

The crustal abundance of a metal also influences the **grade** of ore, the proportion of metal in the mined rock, which is economically viable. The **economic concentration factor** (Fig. 9.2d) is the ratio of this grade to the crustal abundance. Common metals can be profitably mined at low concentration factors (3.5 for aluminium) whereas scarce metals must be much more concentrated (nearly 70 000 for mercury).

Metals have many uses. Steel needs iron, maybe alloyed with titanium, manganese, vanadium, chromium, nickel, tungsten or molybdenum. Zinc and tin provide corrosion-resistant coatings for iron and steel. Aluminium is used in cans, aircraft and for other lightweight components. Electrical applications use copper, zinc, lead, silver, mercury and gold. Titanium and zinc are used in pigments and platinum as a catalyst.

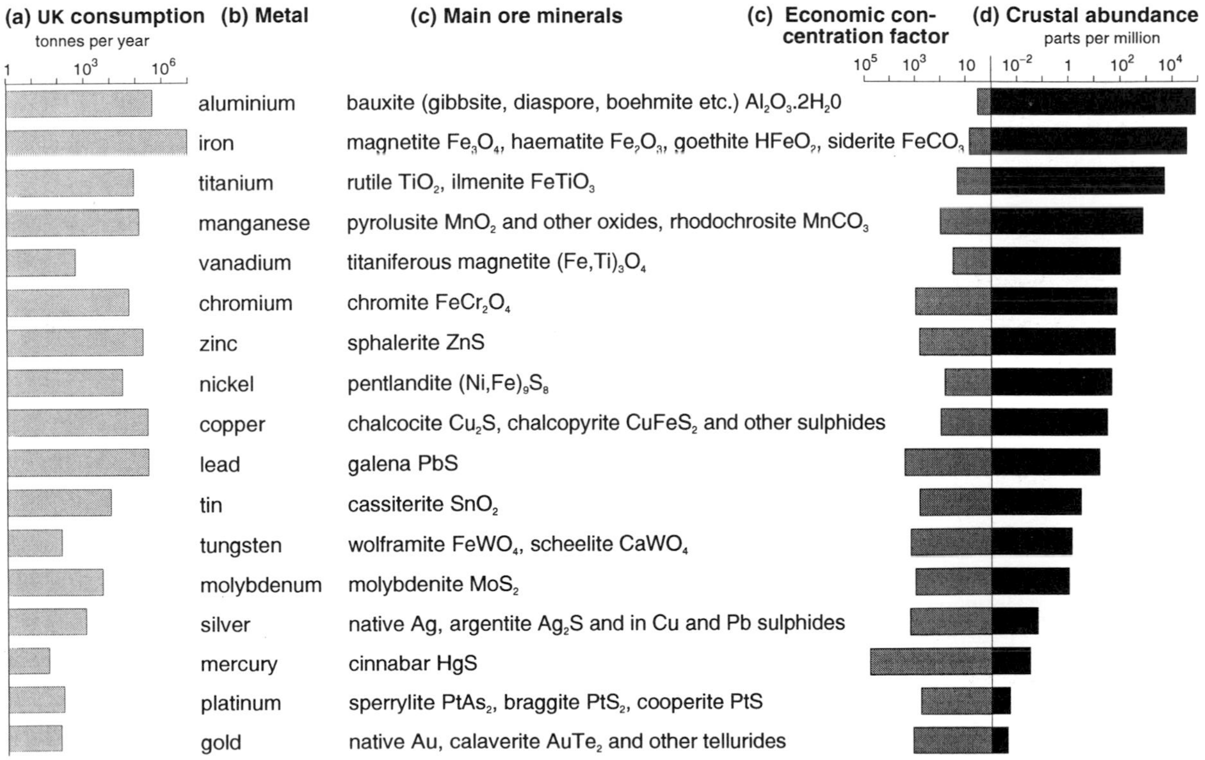

(a) UK consumption (b) Metal	(c) Main ore minerals	(c) Economic concentration factor	(d) Crustal abundance
aluminium	bauxite (gibbsite, diaspore, boehmite etc.) $Al_2O_3.2H_2O$		
iron	magnetite Fe_3O_4, haematite Fe_2O_3, goethite $HFeO_2$, siderite $FeCO_3$		
titanium	rutile TiO_2, ilmenite $FeTiO_3$		
manganese	pyrolusite MnO_2 and other oxides, rhodochrosite $MnCO_3$		
vanadium	titaniferous magnetite $(Fe,Ti)_3O_4$		
chromium	chromite $FeCr_2O_4$		
zinc	sphalerite ZnS		
nickel	pentlandite $(Ni,Fe)_9S_8$		
copper	chalcocite Cu_2S, chalcopyrite $CuFeS_2$ and other sulphides		
lead	galena PbS		
tin	cassiterite SnO_2		
tungsten	wolframite $FeWO_4$, scheelite $CaWO_4$		
molybdenum	molybdenite MoS_2		
silver	native Ag, argentite Ag_2S and in Cu and Pb sulphides		
mercury	cinnabar HgS		
platinum	sperrylite $PtAs_2$, braggite PtS_2, cooperite PtS		
gold	native Au, calaverite $AuTe_2$ and other tellurides		

Figure 9.2 Sime major metals: UK consumption (1991), mineralogy, economic concentration factor and crustal abundance.

9.2 Formation processes

Economic mineral deposits are formed by geological processes that concentrate metals or minerals above their average crustal abundance. These processes are not unusual in themselves, mostly comprising common rock-forming mechanisms (Fig. 9.3). However, they must operate for long enough with the right raw ingredients in a favourable geological environment. Moreover, enrichment in a mineral or metal often occurs through a cycle of successive processes. For example, metallic minerals in low-grade vein deposits may be eroded and further concentrated by sorting in flowing river water. This complexity in forming mineral deposits must be kept in mind as individual processes are described; first the Earth's internal processes, followed by those of the Earth's surface.

Crystal settling in magma chambers causes heavy minerals to accumulate as layers or thin laminae in igneous rocks. Some iron-bearing oxide minerals are dense enough and, equally important, form early enough in the crystallization history to settle in this way. **Magnetite** and **ilmenite** are found in basic igneous rocks, and **chromite** also in ultrabasic rocks.

Immiscibility of liquids in basic magma chambers can concentrate sulphide minerals, notably those of iron, copper and nickel. Sulphide liquids do not mix well with silicate melts, tending to form more dense droplets. These coalesce and sink to the floor of the magma chamber, where they crystallize as pods of ore. **Pentlandite**, **pyrite** and **chalcopyrite** can form in this way, sometimes associated with **platinum minerals** and **gold**.

Metasomatism is the chemical alteration of the country rocks around an igneous intrusion, in addition to any thermal **metamorphism**. Acidic magmas are most effective because they contain most water. Limestones are the most reactive wall rocks. The resulting alteration zone, or **skarn**, is zoned. Oxides such as **magnetite**, **haematite** and **cassiterite** occur close to the igneous contact and sulphides such as **pyrite**, **chalcopyrite**, **galena**, **sphalerite** and **molybdenite** further away, metres or tens of metres from the contact. An accompanying transition occurs in the complex silicate **gangue** minerals.

Intrusion of late-stage magmas can produce coarsely crystalline veins or dykes termed **pegmatites**. These are usually associated with wet acid magmas. Their water and trace elements are not sufficiently accommodated in typical granitic minerals and concentrate in the late magma. Crystallization of this magma in cracks in the chilled granite or country rock yields rare minerals, particularly those bearing lithium, niobium, tantalum, beryllium and zirconium. Tin can be present as **cassiterite**.

Hydrothermal precipitation is often also related to igneous activity, but the mineral components are transported in hot water rather than in magma. The water is either magmatic or reheated groundwater. As this water rises it cools and crystallizes a succession of minerals. Mineral **veins** are tabular deposits formed in discrete fractures, but more diffuse **disseminations** develop by percolation through larger volumes of permeable rocks. These processes are detailed in two examples in the next section (Ch. 9.3).

Leaching is low-temperature chemical alteration of a rock by water to produce a **residual deposit**. Some rock components are dissolved and transported away in solution. Some mineral components suffer **hydroxylation**, the addition of hydroxyl (OH-) ions, to produce clay minerals. The water may be either hydrothermal or surface-derived. Hydrothermal leaching can form **kaolinite (china clay)**, detailed in the next section (Ch. 9.3). Groundwater leaching in tropical climates forms **laterite** soils. These usually contain a mixture of hydrated ferric and aluminium oxides which are difficult to process for either metal. However, where the source rock lacks iron, the valuable aluminium hydroxide **bauxite** can be formed. Leaching of ultrabasic rocks produces laterites rich in insoluble nickel-bearing silicates.

Solution in water concentrates metals not only in the residual deposit but also by *reprecipitation* of the dissolved material. Solution commonly occurs in oxidizing groundwater and reprecipitiation in reducing conditions. In this way low-grade **disseminated** ores can undergo **secondary enrichment** into high-grade pockets. **Copper and silver sulphides** are particularly prone to this, as are **uranium** minerals (Ch. 12.3).

Chemical precipitation of dissolved material occurs from seawater as well as from groundwater. The amount of dissolved oxygen is again important in determining the minerals that form. Fully oxygenated seawater favours deposition of **iron** and **manganese oxides** and **hydroxides**. Water low in oxygen favours **sulphides**, particularly of **copper**, **lead** and **zinc**.

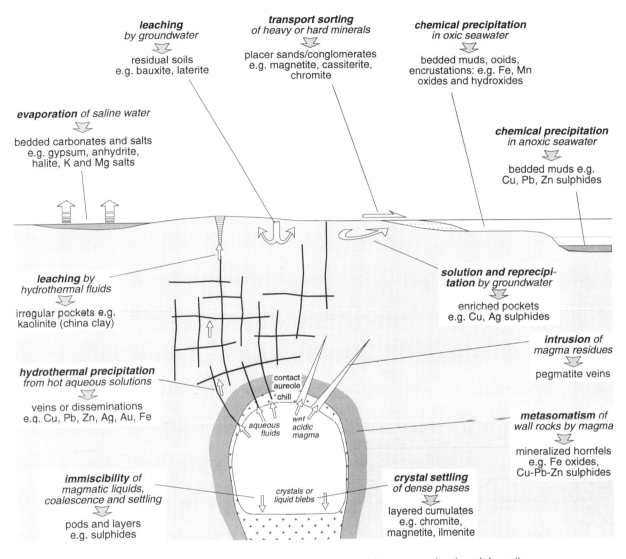

Figure 9.3 Composite cartoon of the processes that form economic mineral deposits.

Evaporation of saline water is one particular mechanism promoting precipitation. It occurs in arid climates when the rate of evaporation of water exceeds its rate of replenishment. The mechanism is detailed in the next section (Ch. 9.3).

Transport sorting operates on eroded sediment in rivers and shallow marine environments, allowing more dense minerals to settle preferentially. Beds enriched in heavy minerals are termed **placers**. Their mineralogy depends on the nature of their source material, but **magnetite**, **ilmenite**, **chromite** and **cassiterite** are common components.

Mineral deposits can be classified into **syngenetic** – those formed at the same time as their host rock – and **epigenetic** – those formed later than their host rock. Processes such as crystal settling, seawater evaporation and transport sorting produce syngenetic deposits, whereas solution/reprecipitation and hydrothermal processes are examples of those producing epigenetic deposits.

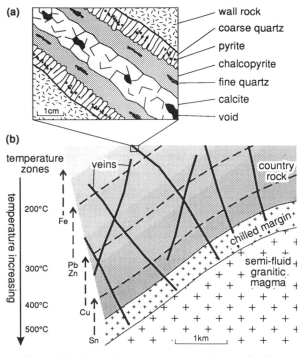

Figure 9.4 Cornish type hydrothermal mineralization.

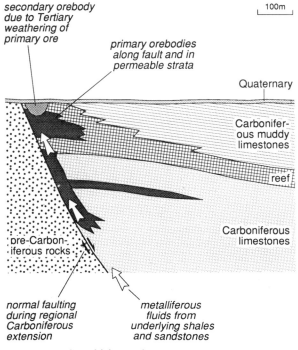

Figure 9.5 Irish type fault-related mineralization.

9.3 Examples of mineralization processes

This section gives four examples of economic mineral deposits in the British Isles that illustrate some common formation processes. The examples can be located both geographically and stratigraphically on Figure 9.8.

Cornish-type hydrothermal mineralization resulted from the intrusion of a granite batholith in southwest England during the Variscan orogeny (Fig. 9.4; Ch. 2.1). The acid magma heated the Devonian and Carboniferous country rocks and began to solidify, partly as a **chilled margin** against the walls and roof of the magma chamber. Water and metals concentrated in the remaining magma. As the country rocks and chilled margin cooled, they developed brittle fractures which allowed the metal-rich fluids to penetrate upwards. Minerals crystallized inwards from the fracture walls to form veins (Fig. 9.4a). A sequence of minerals developed as the fluids cooled during their upward flow, producing a zonation outwards from the granite (Fig. 9.4b). The highest temperature zone is dominated by tin as **cassiterite**, together with **wolframite**, **molybdenite**, **magnetite** and **chalcopyrite**. Chalcopyrite forms a copper-rich zone, before being joined by the lead and zinc minerals, **galena** and **sphalerite**. The low-temperature zone is dominated by the iron sulphide **pyrite**, together with **argentite**, **cinnabar**, and **stibnite**.

Irish-type mineralization is also mainly hydrothermal, but is not related to granites (Fig. 9.5). The mineralizing fluids were probably derived instead from Devonian and early Carboniferous shales and sandstones that underlie the mineralized area. Crustal stretching during Carboniferous time promoted the release of pore water by increasing the regional heat flow. It also formed normal faults in the accumulating Carboniferous rocks, allowing the fluids to permeate upwards. Ore minerals crystallized as veins in the fault zone and as disseminated deposits in permeable limestones above each fault. Zinc and lead minerals predominate, particularly **sphalerite** and **galena**. In some areas a secondary orebody was formed by redeposition and weathering of primary ore when it was exposed at the Tertiary land surface. The secondary deposits include clasts of primary ore in a mudstone formed by oxidation of the sulphide minerals.

China-clay formation in southwest England resulted from hydrous alteration of the same Variscan granites that are associated with the metalliferous mineralization (Fig. 9.6). Potassium feldspar ($KAlSi_3O_8$) reacts with hot acidic water to form the clay mineral kaolinite ($Al_4Si_4O_{10}(OH)_8$). Potassium is removed in solution and some free silica is formed. The original quartz remains unaltered and the mica is either unaltered or is recrystallized in a finer-grained form. The groundwater for kaolinization was probably surface water rather than residual magmatic fluid. Groundwater was heated as it flowed downwards, then returned upwards through fractures in the granite. The heat flow was maintained by the decay of the high proportion of radiogenic minerals in the granite. The resulting kaolinization is concentrated in funnel-shaped zones above major fractures.

Evaporite deposition in northeast England and the North Sea occurred during the arid periods of the Permian and Triassic. At these times the area contained a number of land-locked sedimentary basins which were either non-marine, or had only a narrow marine connection to the open ocean. The arid climate meant that the lakes or shallow seas in these basins suffered rapid evaporation.

The Permian (Zechstein) evaporites illustrate the resulting processes in a marine-influenced basin (Fig. 9.7). As seawater is evaporated, its salinity increases and a sequence of minerals precipitates. **Carbonates** form at slightly raised salinity, then the calcium sulphates, **gypsum** and **anhydrite**. **Halite** forms next, followed at high salinity by salts of **potassium** and **magnesium**. However, this ideal sequence is rarely developed. Inflow of low-salinity water counteracts the effects of evaporation, so that the salinity in the basin can be held stable for a considerable time. The Permian basins alternated between two such stable states. In the **carbonate phase**, near-normal salinities favoured the deposition of limestones at the basin edge (Fig. 9.7a). Only small coastal lagoons could develop salinities high enough to deposit sulphates. In the **sulphate phase**, when evaporated water was replenished more slowly, gypsum, anhydrite and even sodium and potassium salts could be deposited across the whole basin margin. Over the ten million years of late Permian time these two states alternated five times, to give repeated evaporite cycles.

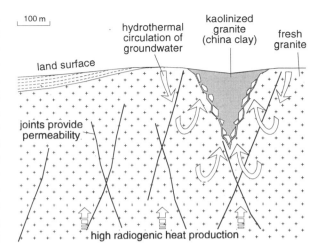

Figure 9.6 Cornish china clay formation.

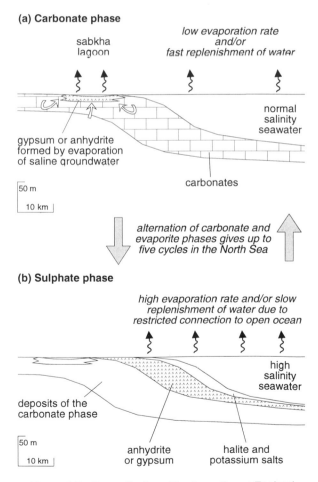

(a) Carbonate phase

(b) Sulphate phase

Figure 9.7 Evaporite deposition in northeast England.

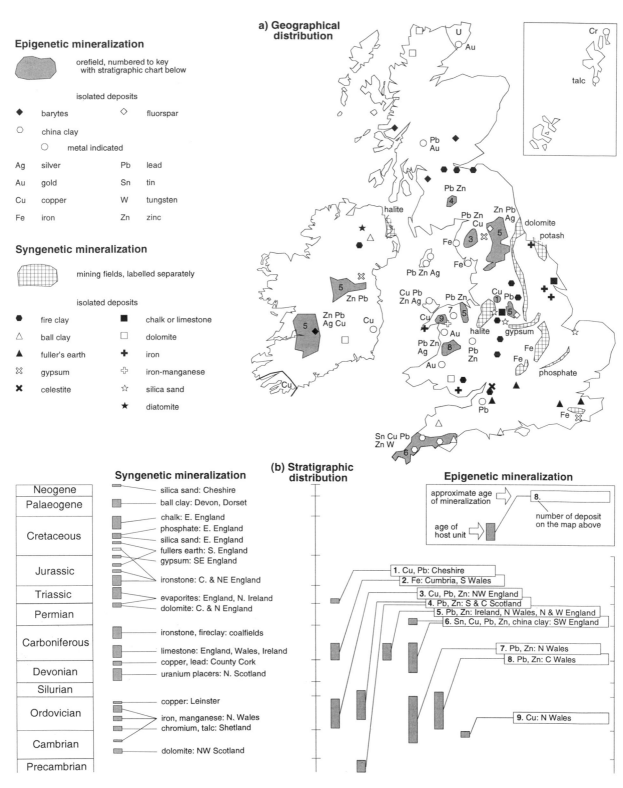

Epigenetic mineralization

orefield, numbered to key with stratigraphic chart below

isolated deposits

◆ barytes ◇ fluorspar

○ china clay

○ metal indicated

Ag silver Pb lead
Au gold Sn tin
Cu copper W tungsten
Fe iron Zn zinc

Syngenetic mineralization

mining fields, labelled separately

isolated deposits

● fire clay ■ chalk or limestone
△ ball clay □ dolomite
▲ fuller's earth ✚ iron
✖ gypsum ✛ iron-manganese
✗ celestite ☆ silica sand
★ diatomite

a) Geographical distribution

(b) Stratigraphic distribution

Syngenetic mineralization

silica sand: Cheshire
ball clay: Devon, Dorset
chalk: E. England
phosphate: E. England
silica sand: E. England
fullers earth: S. England
gypsum: SE England
ironstone: C. & NE England
evaporites: England, N. Ireland
dolomite: C. & N England
ironstone, fireclay: coalfields
limestone: England, Wales, Ireland
copper, lead: County Cork
uranium placers: N. Scotland
copper: Leinster
iron, manganese: N. Wales
chromium, talc: Shetland
dolomite: NW Scotland

| Neogene |
| Palaeogene |
| Cretaceous |
| Jurassic |
| Triassic |
| Permian |
| Carboniferous |
| Devonian |
| Silurian |
| Ordovician |
| Cambrian |
| Precambrian |

Epigenetic mineralization

approximate age of mineralization

age of host unit

8.

number of deposit on the map above

1. Cu, Pb: Cheshire
2. Fe: Cumbria, S Wales
3. Cu, Pb, Zn: NW England
4. Pb, Zn: S & C Scotland
5. Pb, Zn: Ireland, N Wales, N & W England
6. Sn, Cu, Pb, Zn, china clay: SW England
7. Pb, Zn: N Wales
8. Pb, Zn: C Wales
9. Cu: N Wales

Figure 9.8 Geographical and stratigraphic distribution of major mineral deposits in the British Isles.

9.4 Minerals in the British Isles

The British Isles hosts a rich variety of mineral deposits (Fig. 9.8). This section can mention only some of those which are currently worked or have been historically important.

Epigenetic deposits, mostly metallic minerals together with fluorspar or barytes, are restricted to pre-Triassic host rocks (Fig. 9.8b). Epigenetic orefields therefore occur in the north and west of the British Isles (Fig. 9.8a). The deposits can be grouped according to similarities in the age of the host rock and of the mineralization. Common hosts are Lower Palaeozoic sandstones and shales, Devonian sandstones and shales, and particularly Lower Carboniferous limestones. The main mineralization events accompany or follow both the Acadian orogeny, of Devonian age, and the Variscan orogeny, of late Carboniferous to Permian age. Copper mineralization in Cheshire occurred as late as Jurassic time, but since then most crust in the British Isles has been too cool to generate hydrothermal systems. The Palaeogene igneous event in western Scotland has little associated mineralization, probably because neither the magmas nor the country rocks had sufficient metal content.

Syngenetic mineral deposits are particularly abundant in Carboniferous and later rocks (Fig. 9.8b) and are therefore most abundant in England (Fig. 9.8a). The few pre-Carboniferous deposits include the chromium and talc associated with ultrabasic igneous rocks in Shetland, volcanically associated copper deposits in southern Ireland and Anglesey, oolitic ironstones in

Wales and the, as yet unexploited, uranium-bearing sandstones of northern Scotland.

Limestones, although mostly used as construction materials, are sometimes pure enough to be used for more exacting purposes; the Cretaceous Chalk and the Lower Carboniferous limestones are the main sources. The Upper Carboniferous coal-bearing strata contain ironstone and fireclay as well, a fortunate coincidence which resulted in the key geological resources for the Industrial Revolution in Britain. The Carboniferous ironstones were later superseded in importance by the epigenetic hematite ores of northwest England and then by the Jurassic ironstones of central England.

Permo-Triassic evaporites provide the main deposits of halite, potash, gypsum and the strontium sulphate **celestite**, although there are small volumes of Upper Jurassic gypsum in southeast England. Southern England also contains valuable clays. The fuller's earths are volcanically derived bentonites. The china clays of southwest England are altered from Variscan granites before their final cooling (Ch. 9.3), and Palaeogene ball clays their eroded and redeposited equivalents.

Most of the metallic orefields of England and Wales now contain only subeconomic reserves. Production is small compared with their past yields. The lead–zinc orefields of Ireland, discovered only since the 1960s, are much more important (Fig. 9.9b). Recent exploration in the Scottish Highlands has found unexploited economic potential. However, in terms of tonnage and value, production of metallic minerals in the British Isles is now considerably exceeded by that of industrial minerals (Fig. 9.9a).

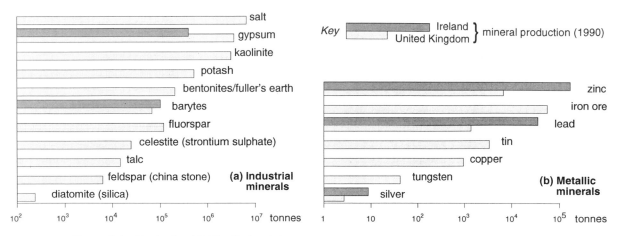

Figure 9.9 Production (1990) of industrial and metallic minerals in the United Kingdom and Ireland.

9.5 Global perspectives

The locally uneven distribution of mineral deposits means that no one country can be self-sufficient in all the commodities that it requires. The shortfall in the UK is seen by comparing production data (Fig. 9.9) with consumption data (Figs 9.1, 9.2). Most industrial minerals are produced domestically in the UK, but phosphates and about half the necessary barytes have to be imported. Surpluses of china clay, ball clay and fuller's earth are exported. Metallic minerals, by contrast, are nearly all imported, with tin the only commodity where recent production has been more than a few per cent of national consumption. Ireland has a much more favourable trade balance in the metallic minerals that it produces.

Much of the industrialized world is similar to the UK in having already exploited most high-grade metallic mineral deposits. The result is a truly international market in metals, their ores and some less common industrial minerals. Global production data show the minerals that are traded in largest volumes (Fig. 9.10). About thirty times as much iron is produced as any

other metal. Most iron is used to make steel, which also absorbs the production of four of the next ten metals; manganese, chromium, nickel and molybdenum. The tonnages given for metallic minerals are in terms of refined metal, whereas those for industrial minerals are as mined (Fig. 9.10). However, the larger volumes of produced industrial minerals are still significant, and show the often underestimated economic importance of this sector of the market.

While production of minerals is spread worldwide, their consumption is heavily concentrated in the rich nations. In the late 1980s, as few as eight industrialized nations consumed about two-thirds of the world's aluminium, copper and lead, over half its iron, and about 60% of its zinc and tin. However, the growth in demand for minerals over the past few decades has not been steady (Fig. 9.11). From 1950 to 1974, growth rates for the more important metallic minerals averaged between 2 and 9% each year, but since then demand has slackened or even declined. Metallic minerals, particularly traditional materials such as lead, tin and iron, have been more affected than industrial minerals, such as kaolinite (Fig. 9.11).

There are several reasons for this pattern. Industrial economies have grown less rapidly since the oil crisis of 1973. They are less involved in heavy industries such as ship-building. Growth industries such as electronics and pharmaceuticals require fewer mineral-

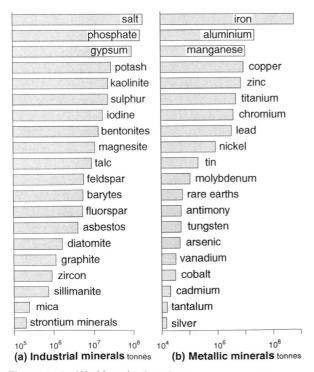

Figure 9.10 World production of major minerals, 1990.

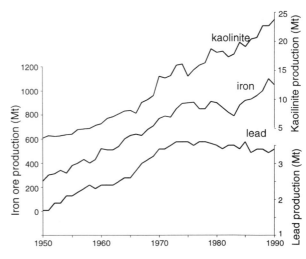

Figure 9.11 Production of some key minerals through time.

derived inputs. New ceramic and plastic materials are replacing metals in many applications. Demand for metal ores has been cut by increased recycling of metals. Recycling is encouraged by the value of metals such as gold and silver, by the toxicity of others such as lead and mercury, and, for aluminium and iron, by the energy saved in refining. Most important, industrialized countries now have much of their infrastructure in place. For these reasons, growth in mineral use is now most rapid in developing countries such as those of South America and southeast Asia.

Estimates of global mineral resources are notoriously difficult to make (Ch. 5). In particular, the "life expectancy" of a mineral, the number of years its reserves or resources would last at current production, must be interpreted with care. However, most estimates show adequate reserves for the immediate future (Fig. 9.12). Only tin, lead, zinc and copper have expected lifetimes of fewer than fifty years.

The exploitation of these remaining mineral resources involves mining ore of increasingly poor grade. Refining low-grade ore requires greater energy input for each tonne of produced metal (Ch. 5). More waste products are produced, increasing the environmental impact of large mines and their associated refining plant (Ch. 13.). These factors, rather than scarcity of mineral deposits themselves, may prove to be more important in constraining future global consumption of minerals.

Further reading

Barnes, J. W. 1988. *Ores and minerals: introducing economic geology*. Milton Keynes: Open University Press. [A sound, readable introduction to the formation and use of minerals.]

British Geological Survey 1992. *World mineral statistics 1986–1990*. Keyworth, Nottingham: British Geological Survey. [A compilation of data on worldwide production and consumption of industrial and metallic minerals.]

British Geological Survey, annually. *United Kingdom minerals yearbook*. Keyworth, Nottingham: British Geological Survey. [Annual compilation of statistical data for the UK together with short reviews of new mineral developments.]

Bowie, S. H. U., A. Kvalheim, H. W. Haslam (eds) 1979. *Mineral deposits of Europe*. Vol. 1, *Northwest Europe*. London: Institute of Mining and Metallurgy. [Includes excellent review chapters on both Ireland and the UK, detailing most significant metallic and industrial mineral deposits.]

Colman, T. B. 1990. *Exploration for metalliferous and related minerals in Britain: a guide*. Keyworth, Nottingham: British Geological Survey. [Short review of UK mineral deposits, concentrating on new finds and future prospects.]

Cruickshank, J. G. & D. N. Wilcock (eds) 1982. *Northern Ireland: environment and natural resources*. Belfast: Queen's University and New University of Ulster. [Contains a review of minerals in Northern Ireland.]

Evans, A. M. 1993. *Ore geology and industrial minerals: an introduction,* 3rd edn. Oxford: Blackwell Scientific. [A reliable and informative explanation of processes with case studies, some from the British Isles.]

Gillmor, D. A. (ed.) 1979. *Irish resources and land use*. Dublin: Institute of Public Administration. [Contains a review of Irish mineral resources.]

Lumsden, G. I. (ed.) 1992. *Geology and the environment in western Europe*. Oxford: Oxford University Press. [Chapter 3 includes a review of mineral resources in a European context together with case studies.]

McLaren, D. J. & B. J. Skinner (eds) 1987. *Resources and world development*. Chichester, England: John Wiley. [Contains valuable reviews of the present and future availability of minerals, as constraints on world development.]

Pattrick, R. A. D. & D. A. Polya 1993. *Mineralization in the British Isles*. London: Chapman & Hall. [Authoritative survey of the main orefields in Britain and Ireland.]

Raistrick, A. 1973. *Industrial archaeology*. London: Paladin. [Contains a short but knowledgeable history of the use of minerals in Britain.]

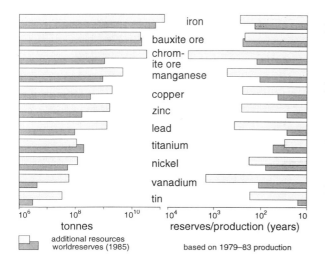

Figure 9.12 World mineral resources and life expectancy.

Chapter Ten

COAL AND PEAT

10.1 Requirements for coal and peat formation

Coal is the compacted and chemically matured form of **peat**, itself a compressed and partially decomposed accumulation of dead land plants. There are four main requirements for coal formation.

Land plants, the main raw ingredient, are found only from late Silurian time (about 425 Ma) onwards, putting a lower stratigraphic limit on the formation of coals. Most major coal deposits formed either in Carboniferous to early Triassic time or between the late Jurassic and Palaeogene. At these periods the global climate, the arrangement of the continents and the evolutionary state of the land flora combined to favour extensive coal-forming swamps.

An *anaerobic environment*, one lacking oxygen, is needed during burial, or else organic matter is quickly oxidized and dispersed after its death. Swamps or bogs are the most common setting, having a water table almost coincident with the sediment surface.

Rapid subsidence of the peat swamp is needed to build up the large thicknesses required to yield economic coal seams. Ten metres of peat compacts to only one metre of coal.

Clastic sediment input is needed to bury the accumulated peat to depths of tens to hundreds of metres where the physical and chemical processes of **coalification** can begin.

10.2 Deltaic coal environment

The first two requirements are met in a range of marginal marine and non-marine environments, giving rise to superficial peat deposits. However, only **deltas** have consistently provided the rates of subsidence and sedimentation sufficient to form thick coal sequences in the geological record.

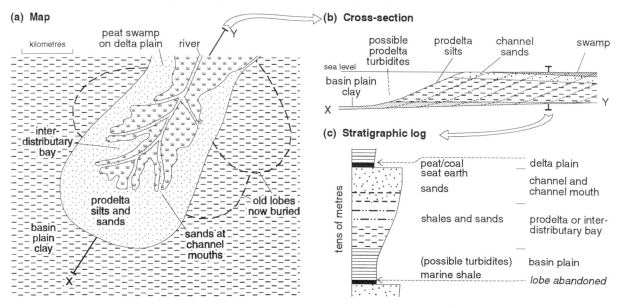

Figure 10.1 An elongate delta viewed on (a) a Map, (b) a vertical cross section, and (c) a stratigraphic log.

Elongate delta model   The formation of deltaic coals is best illustrated by reference to an **elongate or birdsfoot delta** (Fig. 10.1a,b). Such deltas form where coastal waves or tides are too weak to rework river-borne sediment, allowing sediment **lobes** to build out (**prograde**) onto a marine shelf or a lake floor. Coarse sediment, mostly sand, is deposited at the mouths of the delta-top channels, and progressively finer sediment in the **prodelta** region. The prograding lobe leaves in its wake a vertical sequence that coarsens upwards (Fig. 10.1c). The top of the delta lobe develops a branching network of **distributary channels** with **interdistributary bays**. These bays progressively silt up and are colonized by swamp vegetation. The vegetation accumulates as peat which is eventually preserved as a unit near the top of the coarsening-up sequence.

As a lobe advances, its delta-top channels decrease in gradient and become less able to move flood water through the system. Eventually the main channel breaks its banks and finds a shorter route to the sea or lake. The old lobe is abandoned and a new lobe starts to build out to one side of it. The weight of the whole delta on the underlying crust ensures that the abandoned lobe subsides below water level. The lobe is covered first by muds but then by the coarser sediments of another lobe as deposition switches back into the same area. In this way a number of coals can form in one vertical sequence, each recording one episode of lobe advance.

In the Carboniferous of the British Isles, elongate deltas were replaced at times by **sheet deltas**, due to more wave reworking of river-mouth sediment, and by **turbidite-fronted deltas**, which had prodelta sands supplied by density currents.

The earlier coal-bearing sequences of northern England resulted from elongate deltas prograding from the north or east into tens of metres of water (Fig. 10.2a). These Lower Coal Measures show fossil evidence of periodic marine incursions into what was probably a mainly non-marine basin. However, the marine influence declined as the basin was filled with sediment. The Middle and Upper Coal Measures (Fig. 10.2b) were deposited on an extensive delta plain fed by distributary channels that terminated in small lacustrine deltas.

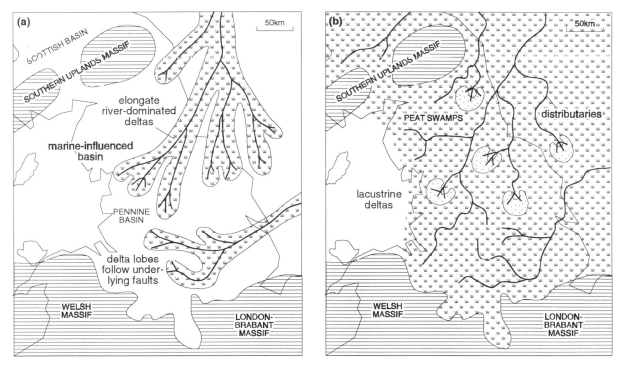

Figure 10.2   Carboniferous environments of central Britain during deposition of (a) Lower and (b) Middle Coal Measures.

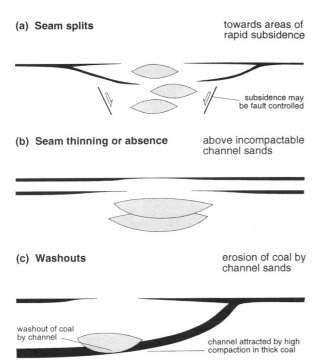

(a) Seam splits — towards areas of rapid subsidence

— subsidence may be fault controlled

(b) Seam thinning or absence — above incompactable channel sands

(c) Washouts — erosion of coal by channel sands

washout of coal by channel —

channel attracted by high compaction in thick coal

Figure 10.3 Subsidence and compaction effects on deltas. Reproduced from *Journal of the Geological Society* (Fielding 1984) with the permission of the Geological Society Publishing House.

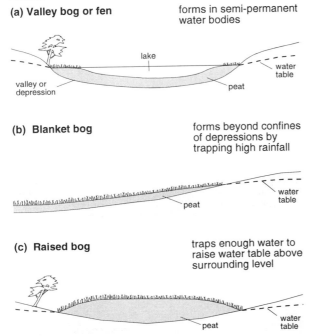

(a) Valley bog or fen — forms in semi-permanent water bodies

lake

water table

valley or depression

peat

(b) Blanket bog — forms beyond confines of depressions by trapping high rainfall

water table

peat

(c) Raised bog — traps enough water to raise water table above surrounding level

peat

water table

Figure 10.4 Peat formation in three types of bog.

Subsidence and compaction on deltas The geometry of deltaic systems is complex. The **progradation** and **switching** of delta lobes give an overlapping mosaic of environments in space and time. Lateral variation in the rate of **tectonic subsidence** is an additional control, causing one area to sink faster than another (Fig. 10.3a). Delta-top channels are repeatedly attracted into the subsiding area. Peat swamps are more persistent outside this area, only intermittently encroaching into it. Resulting coals therefore develop **seam splits** towards the area of rapid subsidence.

Compaction also influences sediment geometry. Peat and mud compact more than sand. One result is that sands in abandoned channels tend to persist as raised ribbons in more rapidly compacting mud or peat swamps (Fig. 10.3b). Resulting coal seams are thin or absent above the old channel. Conversely, a swamp area underlain by thick peat and mud will subside rapidly, forming a topographic low that may attract a newly diverted channel. This channel may erode down into the swamp peat, giving a **washout** in the resulting coal seam (Fig. 10.3c).

10.3 Recent peat deposits

Only some of the Quaternary peat in the British Isles has formed in deltaic or estuarine swamps; the East Anglian fens and the Somerset levels are examples. The largest areas of peat have developed inland, in bogs in river valleys, lake basins or upland hollows.

Three gradational types of inland bog are recognized (Fig. 10.4). **Valley bog** develops next to semi-permanent water bodies, with peat accumulating below water level. **Blanket bog** encroaches beyond valleys or hollows, storing water in its own absorbent organic structure. The most effective vegetation is *Sphagnum* moss, which can trap rain water and grow almost independently of soil nutrients. Vigorous growth of *Sphagnum* can produce **raised bog**, where the water table is elevated above that in the surrounding ground. Peat bogs have accumulated over the 10 000 years since the last glacial maximum, particularly in warmer and wetter periods. They have maximum thicknesses of a few metres. Unlike deltas, these inland environments have no inherent subsidence potential to produce thick deposits.

10.4 Burial and maturation

Biochemical decomposition Dead organic matter is attacked by bacteria in the top few metres of a bog or swamp. The more soluble **cellulose** is progressively broken down, leaving a residue enriched in **waxes** and the woody material **lignin**. The decomposition releases water (H_2O) and gases such as methane (CH_4), carbon dioxide (CO_2) and hydrogen sulphide (H_2S), which usually escape to the atmosphere.

Thermal maturation The transformation of peat into coal occurs by increasing temperature over time, at depths of tens of metres to some kilometres. Changing physicochemical properties define a continuous spectrum of coal maturity or **rank**, divided into **lignite**, **sub-bituminous coal**, **bituminous coal**, **anthracite** and **graphite** (Fig. 10.5). Loss of water and volatiles continues, dominated first by carbon dioxide but later by methane (Fig. 10.6). The proportions first of oxygen and then of hydrogen decrease progressively with respect to carbon (Fig. 10.5). The heat content or **calorific value** of the coal increases correspondingly. Porosity decreases and density increases. The components of higher rank coals acquire a higher **reflectance** to light, a property that can be quantified and used to estimate coal rank.

Coal composition The various classes of organic matter differ in their chemical composition (Fig. 10.6). Woody tissue (**vitrinite**) with its burned or oxidized equivalent (**inertinite**) make up the common **humic coals**. The more unusual **sapropelic coals** are derived from **exinite**, the wax-rich remains of spores and algae. The maturation paths of the different coal types converge at high rank (Fig. 10.6) but their released volatiles differ. Exinite produces more oil whereas vitrinite and inertinite release mainly gas. This distinction will be discussed further in relation to petroleum generation in Chapter 11.2.

Coal contains a variety of impurities, particularly **clay minerals**, **calcium and iron carbonates**, **pyrite** and **sodium chloride**. These components limit the use of certain coals. For instance clay minerals increase the ash residue, sodium chloride accelerates corrosion of boilers and pyrite gives emissions of sulphur dioxide (see Chs 16 & 17).

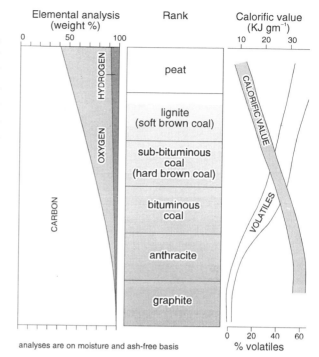

Figure 10.5 Coal properties with increasing rank.

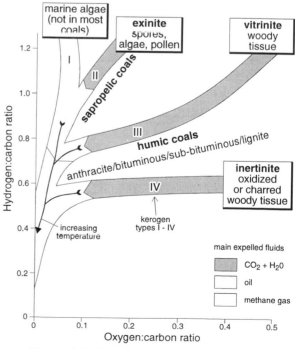

Figure 10.6 The effect of heat on organic matter.

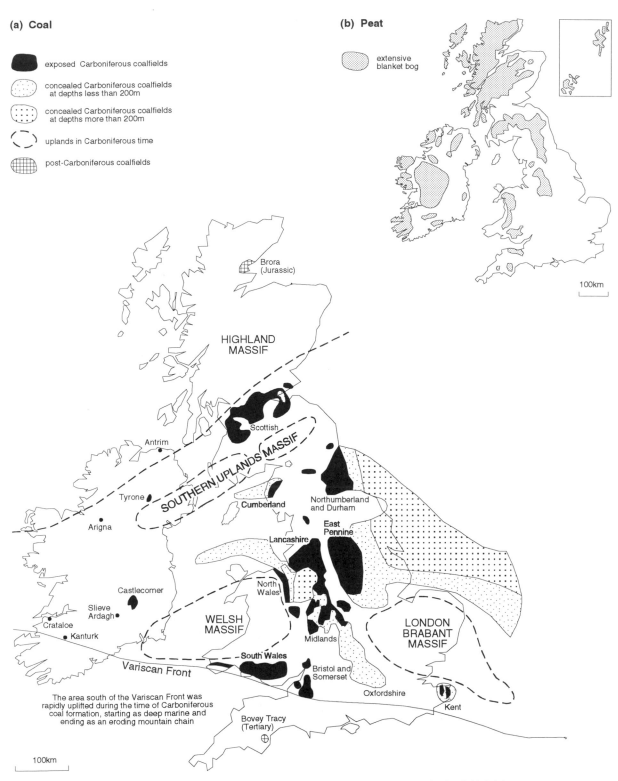

(a) Coal

- ■ exposed Carboniferous coalfields
- ▨ concealed Carboniferous coalfields at depths less than 200m
- ▨ concealed Carboniferous coalfields at depths more than 200m
- ⌇ uplands in Carboniferous time
- ▦ post-Carboniferous coalfields

(b) Peat

- ▨ extensive blanket bog

100km

Brora (Jurassic)

HIGHLAND MASSIF

Scottish

Antrim

SOUTHERN UPLANDS MASSIF

Tyrone

Cumberland

Northumberland and Durham

Arigna

East Pennine

Lancashire

Castlecomer

North Wales

Slieve Ardagh

WELSH MASSIF

LONDON BRABANT MASSIF

Crataloe

Kanturk

Midlands

Variscan Front

South Wales

Bristol and Somerset

Oxfordshire

Kent

The area south of the Variscan Front was rapidly uplifted during the time of Carboniferous coal formation, starting as deep marine and ending as an eroding mountain chain

Bovey Tracy (Tertiary)

100km

Figure 10.7   The location of (a) coalfields and (b) extensive peats in the British Isles.

74

10.5 Geography of coal and peat

Carboniferous coal   The products of the late Carboniferous coal swamps comprise the largest reserve of fossil fuels in the British Isles. Their distribution is strongly controlled by Carboniferous palaeogeography and tectonics (Fig. 10.7a).

The Carboniferous uplands had been created by crustal shortening and uplift during the Caledonian orogeny: the Highland Massif in the Athollian (mid-Ordovician) event and the Southern Uplands, Welsh and London–Brabant Massifs in the Acadian (Early Devonian) event (Fig. 10.7a; Ch. 2). These upland blocks persisted while Early Carboniferous (Dinantian) crustal stretching created marine basins around them. These basins were progressively infilled by deltas in later Carboniferous (Silesian) time. The causes were (a) decreasing basin subsidence as fault-controlled **rifting** gave way to thermal **sag**, (b) increasing southward sediment supply from renewed uplift of the Highland Massif, and (c) the growth of the Variscan mountain belt in the south, destroying marine connections and supplying further clastic sediment.

The **Variscan Front** marks the northern limit of strong Late Carboniferous to Permian deformation, itself localized in the deposits of the formerly deep marine Devonian and Carboniferous basins (Fig. 10.7a). These basins were overwhelmed by deltas for only a brief period before their eventual uplift, so that the Variscan Front also marks the southern limit of productive Carboniferous coalfields.

Most British coalfields have been weakly folded and faulted by the Variscan movements, often by renewed movement on the faults that controlled the Early Carboniferous crustal extension (Fig. 10.8). The fields close to the Variscan Front – South Wales, Bristol and Somerset, Kent – have suffered stronger deformation and hence deeper eventual burial. They contain some

high rank **anthracite**, valued for its high thermal output and low volatile content. However the high frequency of faulting has made these fields difficult to mine, and they have been among the first to be abandoned as demand for coal has decreased in the past three decades (Ch. 10.6).

Coal mining activity naturally began in the **exposed coalfields**, where Upper Carboniferous rocks crop out at the present land surface (Fig. 10.8). However, attention has gradually shifted to the extensive **concealed coalfields** that occur beneath the Permian and Mesozoic rocks deposited after the Variscan orogeny. The totally concealed Kent coalfield was worked from early this century, as were near-surface extensions of exposed coalfields such as in the East Pennine, Northumberland and Durham fields. Major new schemes include the Selby field beneath east Yorkshire, the Vale of Belvoir field in Leicestershire, and the planned development of the Oxfordshire field.

Post-Carboniferous coal   Coal deposits from post-Carboniferous sequences have been worked locally, but these deposits offer very small reserves compared with those of Carboniferous coal (Fig. 10.7a). Examples include those in the Jurassic of Cleveland (Yorkshire) and Brora (Scotland), and in the Tertiary in the Bovey Tracy basin (Devon).

Peat   Extensive blanket peat bogs are restricted to the high rainfall areas of the north and west of the British Isles (Fig. 10.7b). Peat has long been used for domestic fuel wherever it occurs, particularly in Ireland and the Scottish Highlands and Islands. In Ireland, about 60% of cut peat now fuels power-stations, satisfying about 25% of the national electricity demand. A total area of about 60 000 hectares is exploited, yielding about 5 million tonnes of peat per year. The environmental impacts of this large-scale extraction of peat is a major concern (Ch. 14).

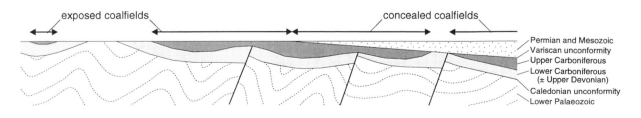

Figure 10.8   Schematic section showing the structural setting of Carboniferous coal.

10.6 Coal consumption and production

Coal is costly to transport, with the result that the UK demand has mostly been satisfied by domestic supplies. Consumption and production trends are therefore closely similar (Fig. 10.9a,b). However, since 1970 coal imports have grown markedly and by 1990 made up about 11% of a UK annual demand of 110 million tonnes. By contrast, the annual demand of about 1.5 million tonnes in the Republic of Ireland must be met mostly by imported coal.

The annual output from UK coalfields grew steadily through the 19th and early 20th centuries, peaking in 1913 at 292 million tonnes. Since then the growing preference for oil and gas as fuels has reduced the need for coal, so that production is now less than a third of its peak tonnage. Most traditional uses of coal have declined or ceased (Fig. 10.9a), so that electricity generation now accounts for three-quarters of UK coal consumption.

The most marked decline in output has been in **deep-mined** coal (Fig. 10.9b). This is cut from underground coal faces in seams 0.5 m–3.0 m thick and is transported to the surface through tunnelled roads and shafts (Ch. 14.3). Mechanization has decreased the unit costs of deep mining, but only for thick seams with simple geological structure. Pit closures have

therefore been most severe in coalfields with complex folding and faulting, such as South Wales, Kent and Scotland.

By contrast, the production of **opencast** coal in the UK has grown steadily (Fig. 10.9b). Coal seams are mined at the land surface after removing unproductive rock overburden (Ch. 14.4). A number of separate seams can be worked at depths of up to 250 m in the same excavation, with the overburden and soil being progressively replaced over the mined-out area. Compared with deep mines, thinner seams can be worked, a higher percentage of each seam can be recovered, and the transport and safety costs are less. Both deep and opencast mining have environmental impacts (Ch. 14).

The proved reserves of coal in the UK are about 9000 million tonnes, about a hundred times the present annual production. However, despite this large reserve relative to oil and gas, coal production looks set to decline even further in the next decade or two. Electricity power stations can currently be fuelled more cheaply by North Sea gas or by coal imported from areas with simpler geology or lower labour costs. Another factor is coal's environmentally dirty image, based on the high emissions of pollutants such as carbon dioxide and sulphur dioxide for each unit of heat derived from it (Chs 16, 17).

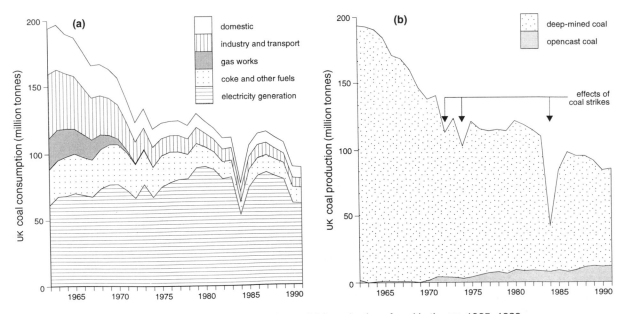

Figure 10.9 Trends in (a) consumption and (b) production of coal in the UK, 1965–1989.

10.7 Global perspectives

World coal consumption rose through most of the 1980s before levelling at around 4.5 billion tonnes per year in the early 1990s (Fig. 10.10a). However, this global pattern hides stronger regional trends. Marked increases in demand in the rapidly industrializing nations of Asia has been balanced by a strong decline in the use of coal in the troubled economies of the former Soviet Union and eastern Europe. Demand in the west European and North American industrial economies has remained comparatively stable.

Most coal is consumed in the region where it is produced. The costs of handling and transporting coal are high relative to its mined value and the range of its uses is less exacting than for oil and gas. North America and non-OECD Europe are the areas with the largest net exports of coal, whereas OECD Europe, Asia and Australasia have the largest net imports.

Another reason for the small international trade in coal is that reserves are spread more evenly around the world than are oil and gas (Fig. 10.10b). Countries with the largest reserves are the USA, the former Soviet Union, China and Australia (Fig. 10.10b). Global reserves would last several hundred years at present consumption levels, a much longer lifetime than for oil and gas. However, the future availability of coal is unlikely to be the main factor limiting its use. Of more immediate concern is the inability of the atmosphere to absorb and recycle the large volume of gaseous pollutants that coal burning produces, particularly the sulphur dioxide that causes acid rain and the carbon dioxide that enhances the greenhouse effect. These constraints will be discussed later (Chs 16, 17).

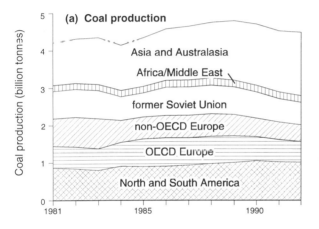

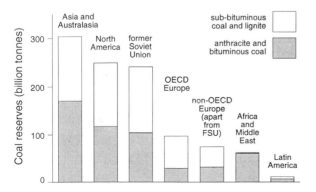

Figure 10.10 World coal production trends (a) and proved reserves (b) by region.

Further reading

Besly, B. M. & G. Kelling (eds) 1988. *Sedimentation in a synorogenic basin complex: the Upper Carboniferous of northwest Europe*. Glasgow: Blackie. [Includes useful reviews of sedimentation and tectonics of British coal basins.]

Brown, G. C. & E. Skipsey 1986. *Energy resources: geology supply and demand*. Milton Keynes: Open University Press. [Chapters 3–5 cover coal, with a British focus.]

Curtis, L. F., F. M. Courtney & S. T. Trudgill 1976. *Soils in the British Isles*. London: Longman. [Includes details of peat formation and distribution.]

Duff, P. McL. D. & A. J. Smith 1992. *The geology of England and Wales*. London: Geological Society. [Chapter 19 has a short review of English and Welsh coalfields.]

Gordon, R. L. 1987. *World coal: economics, policies and prospects*. Cambridge: Cambridge University Press. [A resource-oriented view of the future of coal.]

Thomas, L. 1992. *Handbook of practical coal geology*. Chichester: John Wiley. [An informative manual on finding coal, including a review of global coal sources.]

Trueman, A. E. (ed.) 1954. *The coalfields of Great Britain*. London: Edward Arnold. [The standard reference on British coalfields, useful for those no longer active.]

Ward, C. R. (ed.) 1984. *Coal geology and coal technology*. Oxford: Blackwell Scientific. [Authoritative review of geology and production of coal.]

Chapter Eleven
PETROLEUM

11.1 The importance of petroleum

The twentieth century has been called "the century of oil". Coal, which fired the industrial growth of the previous century, was mostly displaced by this more versatile fuel. In the second half of this century, natural gas has joined oil as a fuel and a feedstock for the plastics, textiles, chemicals and fertilizers on which developed societies rely. But only in the past few decades have the side-effects of this hydrocarbon dependency been assessed: atmospheric smog, marine and river pollution, acid rain and the increase in greenhouse gases.

Oil and gas dominate geology as well as the hydrocarbon society. Half the geological profession is employed in finding them (Ch. 18), and more in researching relevant geological processes. This chapter outlines these processes, some techniques of exploiting oil and gas, and its distribution around the British Isles. The environmental impacts of extracting and using oil and gas are covered in Part IV.

11.2 Requirements for oil and gas

Definitions The term **petroleum** covers a spectrum of natural hydrocarbons. Most important, and the main subjects of this chapter, are **crude oil** and **natural gas** which form the conventional liquid and gaseous parts of this spectrum. The solid component, **asphalt**, is one of several types of **unconventional petroleum**, so called because they are presently uneconomic to extract. However, they form a large hydrocarbon reserve of future economic potential (Ch. 11.7) and possible environmental danger (Chs 13 and 17).

Summary of requirements An economic accumulation of oil and gas results from a coincidence of suitable lithologies, processes and rock geometries (Fig. 11.1). There must be a **source rock** rich in organic matter, petroleum's raw ingredient. The source rock must be heated over time. This **maturation** distils fluid hydrocarbons from the organic matter. **Migra-**

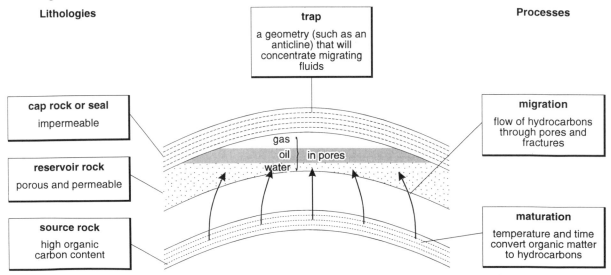

Figure 11.1 Cross-sectional view of the lithologies, geometry and processes needed to form oil and gas.

tion of these fluids out of the source rock is driven first by pressure gradients and then by density contrasts, causing upward flow through pores and fractures. The migrating oil and gas must reach a **reservoir rock** with a high porosity, overlain by an impermeable **caprock** or **seal.** The reservoir and seal must have a geometry or **trap** that concentrates the oil and gas. All these requirements are dependent on the **tectonic history** of the host sedimentary basin.

Source rocks   These need a high organic carbon content. Coals provide carbon in a **concentrated** form and mudrocks in a more **dispersed** form (Fig. 11.2a). Mud is deposited in the calm water also needed for fallout of fine organic particles. The best source-muds form in bottom waters low in oxygen, inhibiting pre-burial oxidation of associated organic matter.

The organic matter in sediments is mostly converted to **kerogen** in the early stages of burial. The nature of the organic material determines the type of kerogen, each with differing proportions of hydrogen, oxygen and carbon (Fig 11.2b). Kerogen type in turn controls the range of hydrocarbons that a source rock can release. In general, terrestrial organic matter gives rise to **gas-prone** source rocks, whereas marine organisms give more **oil-prone** sources.

Maturation   Methane gas is produced by the bacterial decay of organic matter at shallow depths and at temperatures of less than 50°C (Fig. 11.3a). This **biogenic** gas is often lost to the atmosphere, although it may be trapped as frozen **gas hydrate** in the Arctic or the deep sea (see *Traps* below and Section 11.7).

The more economic **thermogenic** gas and oil are produced at temperatures of 50°–200°C (Fig. 11.3a). Higher temperatures favour the release first of oil, then of **wet gas** (so called because it yields liquid condensates on cooling) and last of **dry gas**. Oil will itself break down to wet and dry gas if heated further.

Maturation depends to a lesser extent on time as well as temperature (Fig. 11.3b). The longer the time available for heating, the lower the temperature window in which oil and gas can be generated.

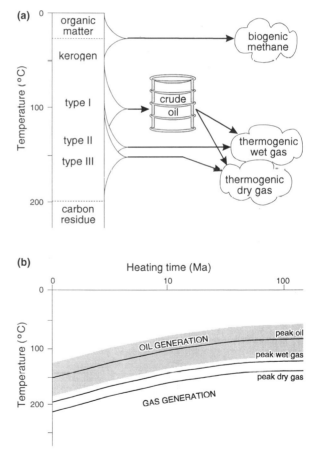

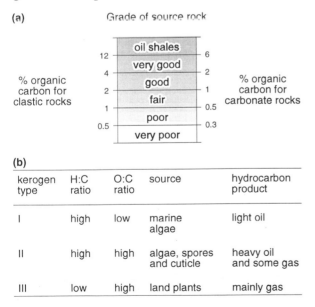

(a)   Grade of source rock

(b)

kerogen type	H:C ratio	O:C ratio	source	hydrocarbon product
I	high	low	marine algae	light oil
II	high	high	algae, spores and cuticle	heavy oil and some gas
III	low	high	land plants	mainly gas

Figure 11.2   The influence on source-rock potential of (a) the percentage of dispersed organic carbon and (b) the type of contained organic matter.

Figure 11.3   The hydrocarbon products of source-rock maturation as a function of (a) temperature and (b) both temperature and time.

Migration   The expulsion of oil and gas out of the source-rock is termed **primary migration** (Fig. 11.4). It is driven by the build-up of pore-fluid pressure in the source rock, owing to either compaction of the pores or the volume increase involved in the maturation of kerogen to hydrocarbon fluids. Being driven by pressure gradients, primary migration can occur both downwards and upwards. However, once expelled into surrounding rocks, **secondary migration** is driven by upward buoyancy forces (Fig. 11.4). The oil and gas percolate up through the water in pore spaces or fractures, and are channelled through zones of high permeability towards a suitable reservoir and trap.

Reservoir rocks   The best reservoir rocks have both a high **porosity** and a high **permeability**. The porosity is the percentage of pore space in the rock, an index of the capacity of the reservoir to hold oil and gas. The permeability is a measure of the ability of the rock to transmit fluids. It depends on how the pore space is interconnected. Permeability is necessary first to allow migration to fill the reservoir with petroleum and then later to permit the extraction of oil and gas through wells drilled into the reservoir.

Porosity arises in three main ways (Fig. 11.5). Most common is the **intergrain porosity** that occurs in most sedimentary rocks. This porosity is reduced by the deposition of crystalline cement. **Fracture porosity** occurs along cracks formed after some cementation of a rock. It may also be reduced by later mineral growth from aqueous solutions. However, some fluids may tend to dissolve rather than precipitate material. In this case a **solution porosity** results from widened fractures and from dissolved rock clasts, particularly carbonate shells.

Caprocks or seals   A caprock must have a very low permeability, in order to impede the upward flow of fluids. Shales seal about two thirds of the world's oil and gasfields, and evaporites most of the remainder. The low permeability of shales is due to their fine grain size and to the close packing of their clay minerals. In evaporites, the interlocking crystalline texture is crucial. Other lithologies may sometimes form seals, including fine-grained fault gouges.

Traps   Oil and gas that accumulated as a thin layer at the top of an extensive horizontal reservoir would be uneconomic to extract. A **trap** is a rock geometry that acts to concentrate fluid hydrocarbons. **Stratigraphic traps** have an inherited depositional or erosional geometry whereas **structural traps** have a geometry created by later deformation.

Examples of stratigraphic traps with depositional geometries (Fig. 11.6a,b,c) include laterally discontinuous sand bodies and carbonate reefs. Unconformities give trap geometries (Fig. 11.6d,e) either by erosional truncation of an underlying reservoir or by deposition of an onlapping reservoir against an underlying basement. Diagenesis can form traps by sealing the updip end of a reservoir by cementation (Fig. 11.6f).

Structural traps are produced by folding or faulting (Fig. 11.6g,h). Such traps usually have a limited storage capacity, related to the amplitude of the fold or to the throw on the fault. If the trap is filled beyond this capacity, the oil and later the gas will leak out laterally at the **spill point** and migrate into an adjacent trap. Folding and faulting can be driven by **diapirism**, the buoyant upwelling of low-density rocks, most commonly salt (Fig. 11.6i).

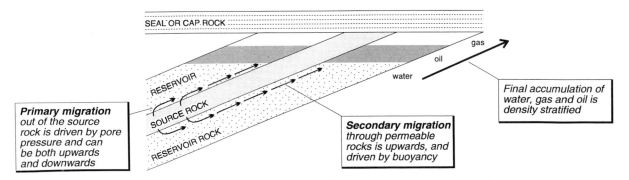

Figure 11.4   Cross-sectional view of primary and secondary migration, and accumulation of oil and gas.

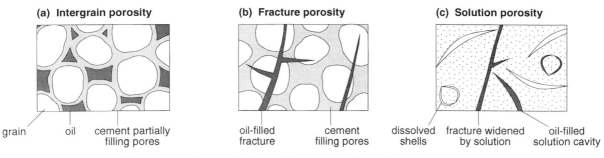

(a) Intergrain porosity	(b) Fracture porosity	(c) Solution porosity
grain oil cement partially filling pores	oil-filled fracture cement filling pores	dissolved shells fracture widened by solution oil-filled solution cavity

Figure 11.5   Microscopic views of three types of porosity in reservoir rocks.

Stratigraphic traps – due to deposition or erosion

(a) Channel fill   **(b) Sediment wedge**   **(c) Reef**

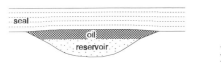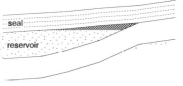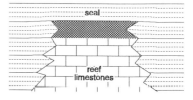

Stratigraphic traps: involving unconformities (d) and (e)   due to diagenesis (f)

(d) Sub-unconformity reservoir   **(e) Supra-unconformity reservoir**   **(f) reservoir with diagenetic seal**

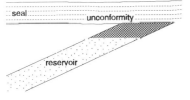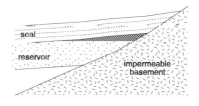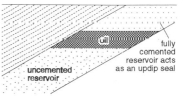

Structural traps – produced by deformation

(g) Anticline   **(h) Fault**   **(i) Salt dome**

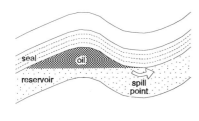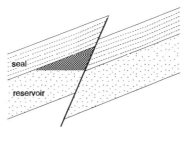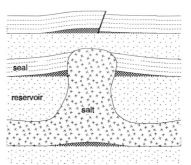

Figure 11.6   Some examples of stratigraphic and structural traps viewed in cross section.

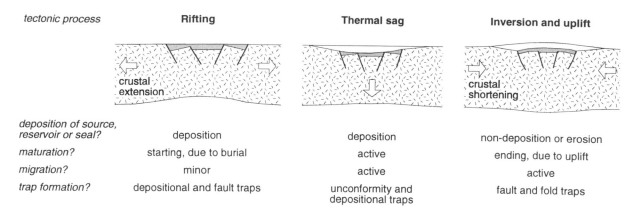

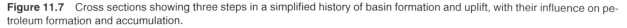

tectonic process	**Rifting**	**Thermal sag**	**Inversion and uplift**
deposition of source, reservoir or seal?	deposition	deposition	non-deposition or erosion
maturation?	starting, due to burial	active	ending, due to uplift
migration?	minor	active	active
trap formation?	depositional and fault traps	unconformity and depositional traps	fault and fold traps

Figure 11.7 Cross sections showing three steps in a simplified history of basin formation and uplift, with their influence on petroleum formation and accumulation.

Control by basin tectonics The formation of oil and gas is regulated by the tectonic history of the host **sedimentary basin** and its underlying crust. One simplified example of a basin history (Fig. 11.7), relevant to the Mesozoic basins around the British Isles involves crustal stretching, to give fault-controlled **rifting** and thermally controlled **sag**, followed by crustal shortening, to give basin uplift or **inversion**. Each stage of such a history can produce different source, reservoir and seal lithologies, a different burial and maturation history and a different combination of traps. Finding oil and gas involves predicting the places in the basin with the right coincidence of all these geological factors. Such a prediction is called a hydrocarbon **play**, a concept illustrated later with respect to the North Sea.

11.3 Exploration and production techniques

Finding oil and gas depends on good geological interpretation, based on data from a variety of exploration techniques. Primary **geological mapping** is carried out only in unexplored onshore areas, although compilation of existing map data is important. **Gravity maps** and **geomagnetic maps** are used in the initial assessment of both onshore and offshore prospects.

Drilling is the most reliable way of establishing subsurface stratigraphic sequences. **Cored holes**, drilled with a cylindrical bit, permit direct lithological logging of retrieved rock. However, they are costly because the whole drill string has to be withdrawn and reinserted to extract each few metres of core. Most holes are therefore drilled with a bit that grinds rocks into millimetre-size cuttings. These are sieved from the drilling mud at the surface and used to reconstruct a lithologic log. **Wireline logs** are produced by lowering a measuring device down the completed hole and electonically recording the response of the wall rocks to a variety of signals. Different lithologies and pore fluids are identified from a combination of responses (Fig. 11.8).

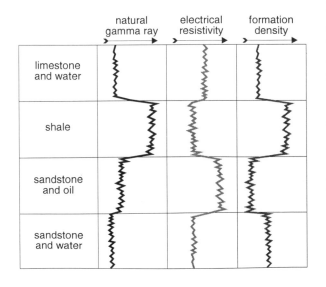

Figure 11.8 Examples of wireline logs through different lithologies.

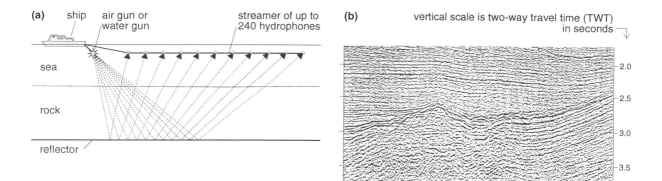

(a) ship air gun or water gun streamer of up to 240 hydrophones

sea

rock

reflector

(b) vertical scale is two-way travel time (TWT) in seconds

— 2.0

— 2.5

— 3.0

— 3.5

Figure 11.9 (a) The principle of seismic reflection profiling at sea. (b) An example of a seismic section with travel time as the vertical axis.

Seismic profiles (Fig. 11.9b) reveal subsurface structure as reflections of energy off stratigraphic and tectonic surfaces. Offshore they are produced by towing a line of **hydrophones** behind a moving ship (Fig. 11.9a). These sensors record the reflected sound waves produced by an air or water gun. The use of multiple recorders increases the number of reflections from each subsurface layer and enhances resolution. Initial data processing produces a **time-section** (Fig. 11.9b), where the vertical scale is the number of seconds to travel down to a reflector and back. Conversion to a more realistic **depth-section** requires a knowledge of the **seismic velocity** of waves through each stratigraphic unit, measured in nearby boreholes.

Geological input to the petroleum industry does not stop when oil and gas are proved by drilling. Precise knowledge of the geometry and permeability of the reservoir are needed to plan the maximum recovery from the field. **Deviated drilling** techniques (Fig. 11.10a) allow an array of holes to penetrate the reservoir from one production platform. The pattern and depth of the holes for **primary recovery** will be influenced by whether oil will be driven up by the pressure of a gas cap above it (Fig. 11.10b) or of the water below it (Fig. 11.10c). Primary recovery usually extracts about 30%–40% of the reserve. A further 40%–50% can be extracted by **secondary recovery** techniques involving injection of either more water or more gas from suitably positioned holes. **Enhanced recovery** relies on reducing the viscosity of the remaining oil by injecting steam, chemicals or even bacteria, but these techniques are not in routine use.

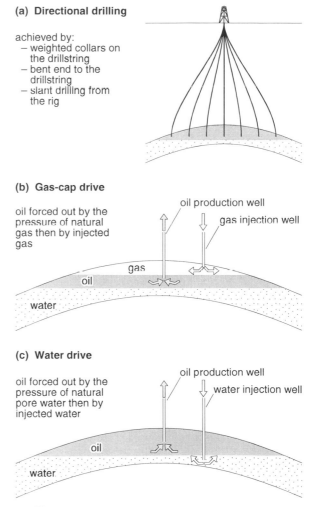

(a) Directional drilling

achieved by:
 – weighted collars on the drillstring
 – bent end to the drillstring
 – slant drilling from the rig

(b) Gas-cap drive

oil forced out by the pressure of natural gas then by injected gas

oil production well

gas injection well

gas

oil

water

(c) Water drive

oil forced out by the pressure of natural pore water then by injected water

oil production well

water injection well

oil

water

Figure 11.10 Oil and gas production techniques.

11.4 Oil and gasfield examples

Geological setting The productive oil and gasfields around the British Isles nearly all occur in the Mesozoic and Cenozoic sedimentary basins that underlie the surrounding shelf seas (Fig. 11.11). These basins resulted from complex Mesozoic crustal extension and thermal sag, similar to that simplified in Figure 11.7. The pattern of fault-bounded linear rift systems can be seen in areas such as the Viking and Central Grabens, the Moray Firth Basin, and the Celtic Sea Basin. Most of the basins have later suffered gentle shortening and resulting uplift, termed **inversion**. However, only the Hampshire Basin and part of the southern North Sea Basin are now exposed onshore, and most remain below sea level.

Four oil and gasfields, three offshore and one onshore, exemplify different combinations of source, reservoir, seal and trap (Fig. 11.12). Each geological combination represents a different hydrocarbon **play**. A more comprehensive view of the main plays around the British Isles is given in Section 11.6 and of the location and size of the resulting fields in Section 11.7.

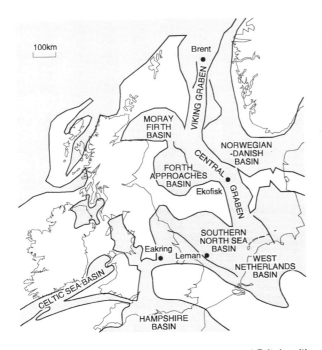

Figure 11.11 Permian to Recent basins around Britain with locations of the oil and gasfields detailed in Figure 11.12.

Brent is one of the largest fields in the Viking Graben of the northern North Sea (Fig. 11.12a). The reservoirs are Jurassic deltaic sheet sands. They have been sourced mainly from the overlying Kimmeridge Clay (Upper Jurassic), first by downward then by lateral migration. The trap geometry was created by Early Cretaceous tilting and faulting. The unconformably draped Upper Cretaceous clays and chalks seal the updip ends of the reservoir sands, which are sealed also by interbedded Jurassic clays. Lower Cretaceous clays seal the faulted eastern end of the lower sand.

Ekofisk is an oilfield in the Norwegian sector of the the Central Graben (Fig. 11.12b). The source rock is the Kimmeridge Clay, as in the Brent field. However, in Ekofisk the Upper Cretaceous seal is fractured and oil has leaked through to be trapped beneath higher Tertiary clays. The reservoir is marine chalk (Upper Cretaceous to Palaeogene). This has a more porous texture than usual owing to soft-sediment slumping. The trap has an anticlinal geometry formed over a rising diapir of Permian salt. This rise also fractured the Upper Cretaceous rocks, providing migration pathways and enhancing the chalk reservoir.

Leman was one of the earliest of the large gasfields to be discovered in the southern North Sea (Fig. 11.12c). Its reservoir is the Rotliegend Sandstone (Lower Permian), a well sorted aeolian sand with high porosity. The overlying Zechstein evaporites (Upper Permian) seal the sandstone and have prevented gas escaping up the many faults which cut the gentle anticline trap. The field has been sourced from the underlying Coal Measures (Upper Carboniferous). Gas production was favoured by the terrestrial source of the organic matter and the high burial temperatures.

Eakring is a small onshore oilfield, economic only because of the low production costs relative to an offshore field (Fig. 11.12d). The reservoirs are channel and channel-mouth sandstones formed on a Late Carboniferous delta. They are sealed by enclosing deltaic clays, and were sourced from marine clays deposited in the pro-delta region. The fault zone next to the field had a normal displacement during sedimentation. The anticline trap was formed when this fault was later reactivated with a reverse sense.

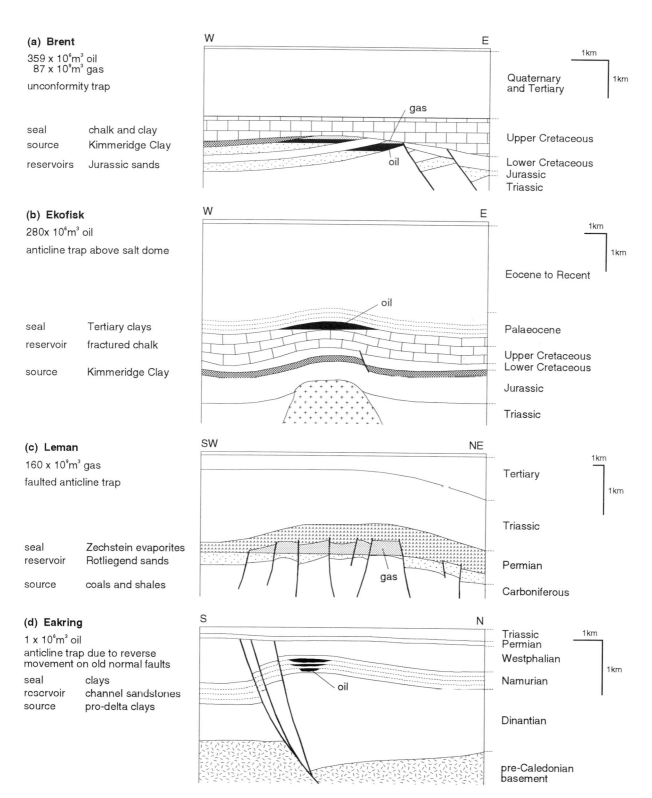

(a) Brent

359 x 10⁶m³ oil
87 x 10⁹m³ gas

unconformity trap

seal	chalk and clay
source	Kimmeridge Clay
reservoirs	Jurassic sands

W   E

gas
oil

Quaternary and Tertiary

Upper Cretaceous

Lower Cretaceous
Jurassic
Triassic

1km
1km

(b) Ekofisk

280x 10⁶m³ oil

anticline trap above salt dome

seal	Tertiary clays
reservoir	fractured chalk
source	Kimmeridge Clay

W   E

oil

Eocene to Recent

Palaeocene

Upper Cretaceous
Lower Cretaceous

Jurassic

Triassic

1km
1km

(c) Leman

160 x 10⁹m³ gas

faulted anticline trap

seal	Zechstein evaporites
reservoir	Rotliegend sands
source	coals and shales

SW   NE

gas

Tertiary

Triassic

Permian

Carboniferous

1km
1km

(d) Eakring

1 x 10⁶m³ oil

anticline trap due to reverse movement on old normal faults

seal	clays
reservoir	channel sandstones
source	pro-delta clays

S   N

oil

Triassic
Permian
Westphalian

Namurian

Dinantian

pre-Caledonian basement

1km
1km

Figure 11.12   Examples of oil and gasfields around the British Isles, located on Figure 11.11.

11.5 Hydrocarbon plays

The fields in Chapter 11.4 exemplify only four of at least eight important hydrocarbon plays around the British Isles. Each play involves a particular combination of source, reservoir and caprocks (Fig. 11.13). A productive play also needs the right history of maturation, migration and trap formation, realized only in local areas of any basin. In the North Sea, productive plays cluster in the main rift zones (Fig. 11.14). Here, deeper burial and higher heat flow have matured the source rocks, and the rift deformation has promoted structural and stratigraphic traps.

The hydrocarbon plays are described in the context of the regional geological history. An overview of this history is given in Chapter 2.

Carboniferous   The late Carboniferous deltas that inundated the southern British Isles and North Sea produced the area's oldest productive sources and reservoirs. The deltas were fed by sand and mud from the growing Variscan mountains in the south and from rejuvenated Caledonian mountains in the north. Humid equatorial conditions promoted abundant delta-top vegetation and resulting coals. These source the major gasfields of the southern North Sea (plays 2,3). Pro-delta muds provide an oil-prone source for small onshore fields reservoired in local delta-top or pro-delta sand bodies (play 1, e.g. Eakring; Fig. 11.12d).

Permian and Triassic   The crustal compression of the Variscan orogeny gave way to extension in the early Permian. This rifting initiated the important Mesozoic basins (Fig. 11.11), though mostly along old structural trends in the basement. Northward continental drift into arid latitudes encouraged deserts to form on the old delta plains. The windblown Rotliegend sands (early Permian) form a major gas reservoir in the southern North Sea Basin (play 2), though not further north where the underlying Carboniferous source rocks are absent. The overlying Zechstein evaporites, precipitated in late Permian desert lakes, form a wide-

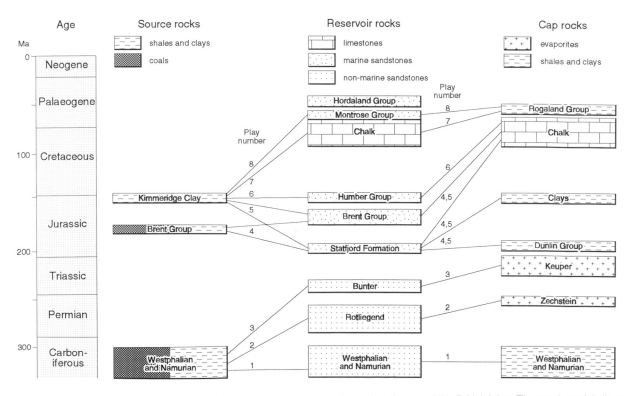

Figure 11.13   Stratigraphic distribution of the main sources, reservoirs and seals around the British Isles. The numbered tie lines indicate hydrocarbon plays; their geographical extent is shown on Figure 11.14.

spread caprock (e.g. Leman field; Fig. 11.12c).

Arid conditions continued into Triassic time, first providing the Bunter sandstone reservoir and then a succession of evaporite seals, particularly those of the Keuper Formation (play 3). The Triassic gasfields are common in areas where the Permian evaporite seal was not deposited or was breached by later faulting.

Jurassic and Early Cretaceous   Renewed late Triassic to Jurassic rifting coupled with globally rising sea levels promoted the marine-influenced basins of the Jurassic and Cretaceous. Coastal, deltaic or shallow marine sands provide reservoirs at three main stratigraphic levels (plays 4,5,6; Fig. 11.13). Coals and shales of the Brent Group (Middle Jurassic) provide a gas-prone source (play 4) but the main oil-prone source for Jurassic and later reservoirs is the Kimmeridge Clay (Upper Jurassic). This source matured only in rapidly subsiding rifts such as the Viking and Central Grabens, giving a pattern of plays that reflects the underlying structure (Fig. 11.14). The Kimmeridge

and lower marine clays provide effective local caprocks for the Jurassic reservoirs.

Late Cretaceous to recent   A mid-Cretaceous change from fault-influenced subsidence to thermal sag (simplified on Fig. 11.7) is associated with a marked unconformity. The overlying Upper Cretaceous marls seal the updip ends of any eroded Jurassic reservoirs (e.g. Brent field; Fig. 11.12a). The Chalk forms a reservoir only where its porosity is enhanced by fracture or slumping (play 7, e.g. Ekofisk; Fig. 11.12b). Gentle late Cretaceous to Palaeogene compression generated new structural traps, but also ruptured some gas traps in the southern North Sea.

Continued Palaeogene to Recent sag centred on the Central and Viking Grabens localized turbidite reservoir sands in the lower Palaeogene (play 8). The enduring subsidence was crucial in maturing the main Kimmeridge Clay source. Oil and gas generation began in the graben centres in late Cretaceous to Palaeogene time and continues today.

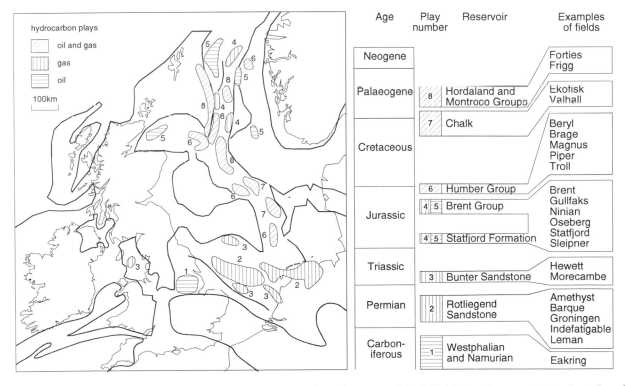

Figure 11.14   Geographical distribution of the main hydrocarbon plays around the British Isles, shown on the basin outline of Figure 11.11. Examples from the major fields are arranged in stratigraphic order of reservoir rocks.

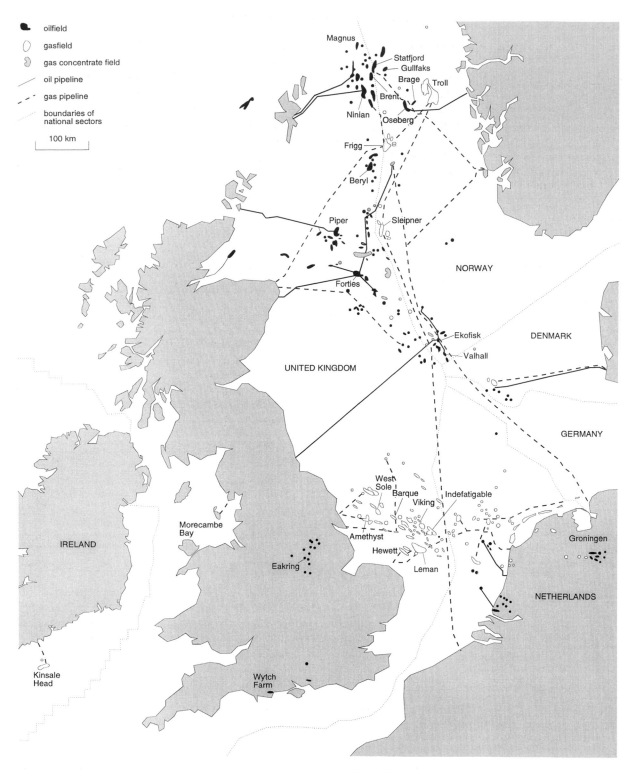

Figure 11.15 Location of oil and gasfields and major pipelines around the British Isles. The twenty largest fields are named (see Fig. 11.16) along with selected other fields only.

11.6 Production in space and time

The geography of oil and gasfields around the British Isles (Fig. 11.15) reflects the pattern of geologically controlled hydrocarbon plays (Fig. 11.14). Successive plays have been explored over the past thirty years.

The first prospects were the Permian and Triassic gas plays of the southern North Sea encouraged by the discovery in 1959 of the giant Groningen gasfield, still the second largest in the region (Fig. 11.16). West Sole and Viking were the first finds in the British sector (1965), followed by the larger Barque, Leman, Indefatigable and Hewett fields (1966). The exploration effort expanded north into the Central Graben, targeting gas reservoirs in Palaeogene sands but first finding unexpected oil in the Chalk at Valhall (1967) and Ekofisk (1969). The Palaeogene play eventually paid off, for example with oil at Forties (1970) and with gas at Frigg (1972).

By 1971 exploration had spread still further north to Jurassic reservoirs in the Viking Graben. Brent was the first major find, followed by many others such as Beryl and Piper (1972), Ninian (1973), Magnus, Statfjord and Sleipner (1974). New gas finds continued in the Permo-Triassic reservoirs of the North Sea, the largest being Amethyst (1972). The Triassic also yielded the Morecambe gasfield (1974). It and the Kinsale Head gasfield (1971) are the only two significant finds to the west of mainland Britain.

The late 1970s saw major discoveries in the Jurassic of the Norwegian North Sea, first Gullfaks (1978) then Troll and Oseberg (1979). The last two and Brage (1980) marked successful exploration on the eastern flank of the Viking graben. During the 1980s and early 1990s the discovery rate of oil and gas slowed to less than half that in the 1970s. Future finds are most likely to be small satellites around existing production fields.

The discovery history of oil and gas around Britain is reflected in the statistics for production and consumption through time (Fig. 11.17). Consumption of natural gas in both the UK and Ireland nearly follows its production offshore. The apparent UK production shortfall since 1977 is imported gas, mostly from the Norwegian North Sea. By contrast, the vigorous import and export of different grades of crude oil mean that UK oil production is unrelated to changes in domestic consumption. A net surplus of oil since 1980 has been decreasing since 1986 as production declines (Fig. 11.17).

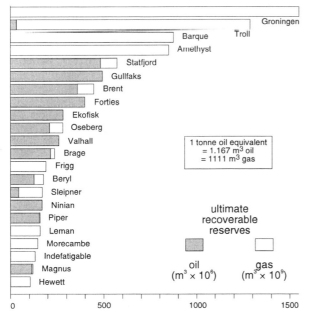
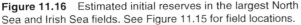

Figure 11.16 Estimated initial reserves in the largest North Sea and Irish Sea fields. See Figure 11.15 for field locations.

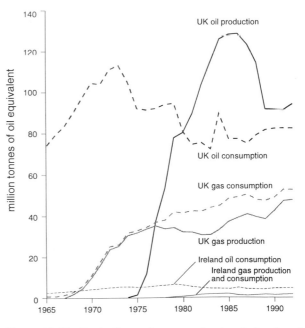

Figure 11.17 Production and consumption trends for oil and gas in the UK and Ireland.

11.7   Unconventional petroleum

Heavy and extra heavy oil has a viscosity higher than that of conventional crude oil and requires special methods of extraction, transport and refinement. About 5% of heavy oils are immature hydrocarbons that have not migrated from their source rock; most have been matured, migrated and trapped in the same way as conventional oil, but have been altered later in the reservoir. The alteration processes include water washing, bacterial degradation and evaporation, resulting in increases of viscosity, specific gravity, and the sulphur and trace metal content.

Steam injection into the reservoir is used to lower the viscosity of heavy oil before it can be pumped out. Between a quarter and a third of the energy equivalent of the extracted oil goes in generating this steam, so that only rich accumulations of heavy oils are economic. The Orinoco Oil Belt of Venezuela contains over half the world's reserves (Fig. 11.22). Some British electricity generating stations are now burning Orimulsion, an emulsion of water and Orinoco heavy oil.

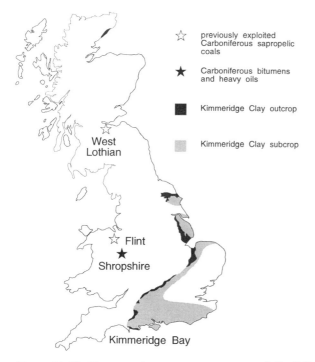

Figure 11.18   Unconventional petroleum resources in the British Isles.

Bitumen has a similar origin to heavy oils but has an even higher viscosity. It acts as a cement in the pores of the reservoir rock, forming **tar sands**. Injected steam or gas may be used to mobilize the bitumen in deeply buried tar sands. However most economic exploitation of tar sands, notably the Athabasca Tar Sands of Canada, is by surface mining followed by heating in a retorting plant.

The first petroleum to be exploited in Britain, by the Romans and later during the 17th century, was bitumen and heavy oil from Carboniferous sandstones and breccias in Shropshire (Fig. 11.18).

Shale oil is a product derived from kerogen in organic-rich rocks, typically **oil shales** or less commonly **sapropelic coals**. Suitable rocks have been buried to only shallow depths, so that their kerogen has not matured or migrated and has remained in its source rock. The rock is usually mined and the oil distilled from it by heating to about 450°C. The energy required for this extraction is equivalent to about 40 litres of oil per tonne of rock, giving the minimum yield for an economic oil shale. This corresponds to about 7% by weight of organic matter. Alternative methods of utilizing shale oil are by *in situ* distillation, and by burning of raw shale in power stations. Only a small fraction of the vast world reserves of shale oil are presently an economic resource using any of these methods.

Sapropelic or **cannel coal** from the Carboniferous of Scotland was exploited for at least a century until 1962 (Fig. 11.18). Such coals are thinner and less persistent in southern Britain. Here the richest oil shales are within the Kimmeridge Clay Formation (Upper Jurassic; Fig. 11.18). Although most of the formation has higher-than-average kerogen contents, these rise to between 7% and 40% in five discrete seams that might just be economic in the future.

Unconventional gas is held in deeply buried or low permeability reservoirs that are not amenable to normal methods of extraction. **Geopressurized natural gas** is dissolved in the pore water rather than forming an immiscible gas cap. This solution occurs at depths of between 5 km and 15 km. Here drilling is costly or impossible and the pore fluids are hot and often very saline, making most deep gas economically unattractive at present. High extraction costs are also associated with **tight-reservoired gas**. This is held in low permeability sandstones and shales which would need to be explosively or hydraulically fractured before gas could be extracted. The major resources in the British Isles are likely to be in Carboniferous sandstones. A special type of tight-reservoired gas is **coal-bed methane**. About 100 million cubic metres of this are extracted each year from British coal mines, to reduce the risk of explosion, and used for local heating. Drilling for coal-bed methane as a resource in its own right is a future possibility.

Methane clathrate is a frozen form of methane in water, also known as **gas hydrate**. Unlike ice, it is stable at temperatures above zero if buried at a few hundred metres or more (Fig. 11.19). Clathrates are stable in two natural environments. Low temperatures promote their formation beneath ice sheets or permafrost (geothermal gradient in Fig. 11.19). The high water pressure beneath the oceans promotes freezing in continental margin sediments. In both cases, the trapped methane is shallow biogenic or deeper thermogenic gas that would otherwise escape upwards. Future commercial exploitation of clathrates would be by heating or depressurization. A more immediate concern is of uncontrolled melting of clathrates due to rising global temperatures (Ch. 17).

11.8   Uses of petroleum

Crude oil is **refined** by **fractional distillation**. The crude is heated to about 360°C and pumped into the base of a fractionating tower. Liquids with successively lower boiling points condense as they rise higher in the tower. The resulting **fractions** range from highly viscous **wax** and **bitumen** to gaseous **liquid petroleum gas** (Fig. 11.20). Waxes and heavy oils can be further converted to petrols by **cracking**, using zeolite catalysts or hydrogen. In this way the proportion of each fraction can be varied to meet local demand. In wealthy countries the proportion of gasoline used is higher than the global average, up to half the total in the USA. In less developed countries fuel oil and gas oil are proportionally more important.

Whatever the country, about 90% of the refined oil and the natural gas is burned as a fuel. The emissions from this combustion – carbon dioxide and monoxide, nitrous oxides, sulphur dioxide and unburned hydrocarbons – have a major atmospheric impact (Ch. 17). The remaining 10% of petroleum products go as a feedstock to the chemical industry and a variety of end products (Fig. 11.20).

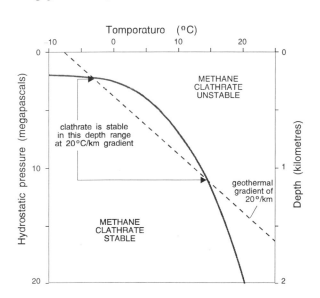

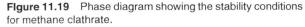

Figure 11.19   Phase diagram showing the stability conditions for methane clathrate.

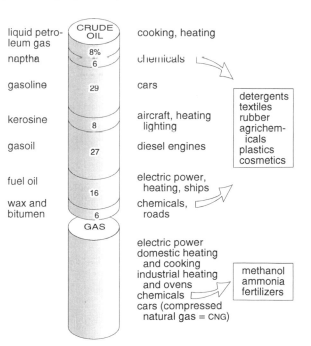

Figure 11.20   End uses of natural gas and fractions of refined crude oil. Percentages are world averages.

11.9 Global perspectives

Oil and gas are transnational commodities. Their exploration, production and supply are dominated by multinational companies. Their price is controlled by international economics and politics. Their use creates atmospheric wastes that have global impacts.

Petroleum production and consumption   Through the first three-quarters of the twentieth century the consumption of oil roughly doubled in each decade. If this exponential growth had continued, the world in 1990 would have been using about 10 billion tonnes (tonnes $\times 10^9$) each year, about the remaining resources of the whole of western Europe. In fact, oil consumption in 1990 was about 3.1 billion tonnes. Its rise was curbed first by the Middle East oil embargo of 1973 and then by the effects of the Iranian crisis in 1979–80 (Fig. 11.21b). But since 1985 consumption has risen steadily again, in response to relaxation of fuel efficiency controls and to optimism about oil reserves outside the Middle East.

Natural gas production has been doubling about every twenty years. It has fluctuated less in response to political and economic pressures, because of its more even geographic distribution and the lower volume of its international trade (Fig. 11.21a).

Petroleum reserves   Some of the problems of estimating reserves of geological resources have already been discussed (Ch. 5). The amount of **proved reserves** (Fig. 11.22) increases as new fields are discovered and as recovery techniques from the reservoirs are improved, but is partially or wholly offset by increasing **production**. Global reserves of oil would last 43 years and of gas 58 years at present rates of consumption. The estimates of **unproved resources** vary with increased geological understanding of petroleum provinces. These assumed resources add a further 23 and 75 years to the lifetimes of oil and gas respectively. Reserves of heavy oil and bitumen are also substantial, but most are currently sub-economic. Other unconventional petroleum reserves will be discussed in their global context in Chapter 12.

The estimates of reserves show the markedly different global distributions of each petroleum component (Fig. 11.22). Oil is hugely concentrated in the Middle East, whereas gas is also abundant further north in the republics of the former Soviet Union. By contrast, the major known reserves of heavy oil and bitumen are in South and North America respectively. If the historical precedents of natural resource conflicts (Ch. 3) are any guide, this geography of petroleum is likely to play a major rôle in the international politics of coming decades.

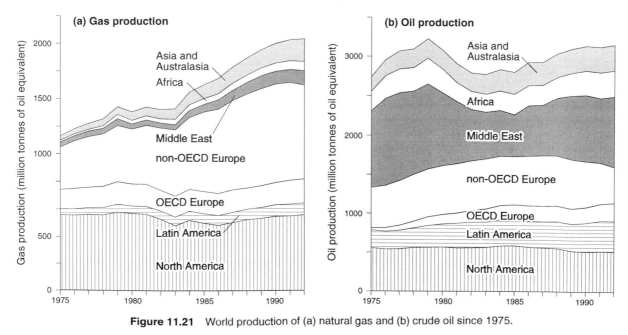

Figure 11.21   World production of (a) natural gas and (b) crude oil since 1975.

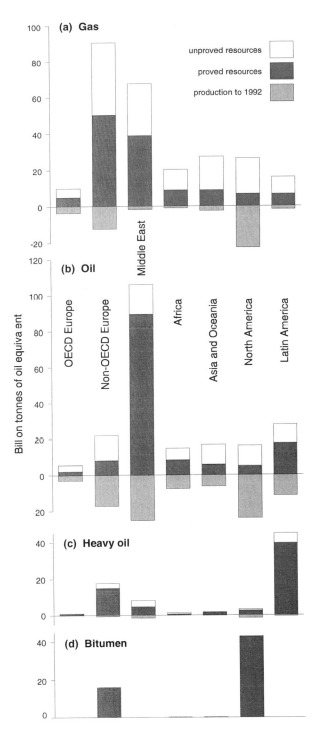

(a) Gas

unproved resources
proved resources
production to 1992

(b) Oil

Bill on tonnes of oil equiva ent

Middle East
OECD Europe
Non-OECD Europe
Africa
Asia and Oceania
North America
Latin America

(c) Heavy oil

(d) Bitumen

Further reading

Abbotts, I. L. (ed.) 1991. *United Kingdom oil and gas fields: 25 years commemorative volume.* Memoir 14. London: Geological Society. [Standardized descriptions of the oil and gasfields in the UK offshore provinces.]

Allen, P. A. & J. R. Allen 1990. *Basin analysis: principles and applications.* Oxford: Blackwell Scientific. [Chapters 10 and 11 give a clear detailed review of petroleum play principles.]

British Petroleum 1993. BP *statistical review of world energy 1993.* London: British Petroleum Company plc. [Annually updated information on the production, consumption and reserves of fossil fuels.]

Brown, G. C. & E. Skipsey 1986. *Energy resources: geology supply and demand.* Milton Keynes, Bucks.: Open University Press. [Chapters 6–9 cover oil and gas, with a British focus.]

Department of Energy 1993. *Development of the oil and gas resources of the United Kingdom.* London: HMSO. [The "Brown Book"; an annual review of petroleum activity.]

Geological Museum 1988. *Britain's offshore oil and gas.* London: British Museum (Natural History). [A short but accurate introduction with excellent graphics.]

Glennie, K. W. (ed.) 1990. *Introduction to the petroleum geology of the North Sea,* 3rd edn. Oxford: Blackwell Scientific. [A stratigraphically arranged review of the whole North Sea, with a useful chapter on exploration history.]

McQuillin, R., M. Bacon, W. Barclay 1984. *An introduction to seismic interpretation: reflection seismics in petroleum exploration,* 2nd edn. London: Graham & Trotman. [A basic guide to seismic methods.]

North, F. K. 1985. *Petroleum geology,* 2nd impression. Winchester, Mass.: Unwin Hyman. [A comprehensive manual on finding oil and gas, mostly with North American examples.]

Selley, R. C. 1988. *Applied sedimentology.* London: Academic Press. [Sedimentological principles, with a strong emphasis on those relevant to finding petroleum.]

Yergin, D. 1991. *The prize.* New York: Simon & Schuster. [Definitive history of the oil industry worldwide.]

Chapter Twelve
ENERGY PERSPECTIVES

12.1 Energy use

The intensive 20th-century search for fossil fuels has been driven by a growth in global energy demand that was exponential until the oil crises of 1973 and 1979 (Fig. 12.1a). Consumption continues to increase, by over 50% in the two decades to 1990. This chapter assesses the potential of fossil fuel reserves to supply future energy demands. It then reviews alternative energy sources, particularly those involving geological factors, and their abundance in the British Isles.

The world trend in energy use hides some important regional variations. The curves for the UK and Ireland (Fig. 12.1b) are typical of Western industrialized nations. Total energy use rose sharply through the 1950s and 1960s with oil use expanding at the expense of coal. Since 1973, total use has fluctuated but has not grown, despite a large increase in the use of natural gas. This fragile stability is due to the decline of heavy industry and to energy efficiencies in response to increased fuel costs.

World growth in energy use was maintained during the 1970s and 1980s by two groups of countries: those of the former Soviet Union and Central Europe, and the rapidly industrializing countries, such as Brazil, Malaysia and Mexico. The first region was self-sufficient in fossil fuels, immune from the 1970s oil shocks and energy efficiency pressure. By contrast, rapidly increasing energy use in industrializing countries is due to increasing economic output as well as low energy efficiency.

Despite high energy growth in developing countries, most world energy is still used by rich nations. There is a strong correlation between a nation's per capita fossil fuel use and Gross National Product (Fig. 12.2). Energy-efficient nations, such as Switzerland and Japan, plot below the general trend and inefficient nations, such as China and the Gulf oil states, plot above. The seven largest OECD economies between them use about 43% of world fossil fuel production, despite having only 12% of its population. Future global fossil fuel needs therefore depend as much on continued high demand in the rich world as on demand growth in the developing world.

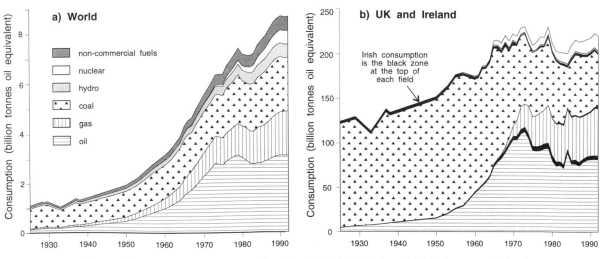

Figure 12.1 Total energy consumption since 1925 (a) globally and (b) in the UK and Ireland.

94

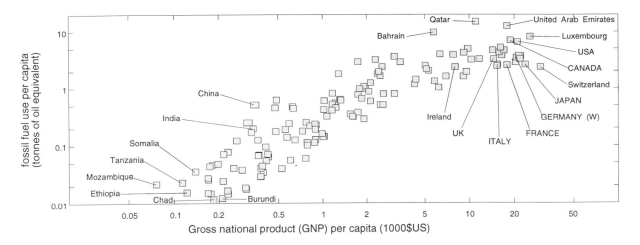

Figure 12.2 Fossil fuel consumption plotted against wealth. The seven largest OECD countries are in upper case.

12.2 Fossil fuel reserves and shortfall

Is there a large enough fossil fuel reserve to meet this growing energy demand into the future? The proved reserves of fossil fuels equate to about 1386 billion tonnes of oil (Fig. 12.3a). At the 1990 global energy need of about 8 billion tonnes, this reserve would last about 170 years. The estimated unproved resources of 6815 billion tonnes would add another 850 years to these totals. These are deceptively large numbers. Much of the proved reserves is shale oil, heavy oil and bitumen, the extraction and refining of which is itself an energy-intensive process. This reduction in useful energy applies also to the unproved resources, most of which will be in geologically and technologically difficult situations. Only a fraction of this resource may be exploitable.

Prediction of future energy supply and demand is very hazardous, because the uncertainties about economic projections are even greater than those about energy resources. However, most models predict that fossil fuels will be insufficient to supply world energy needs through the next century (Fig. 12.4). This is even on the unlikely assumption that no constraints are imposed on fossil fuel use because of the atmospheric pollution that it causes (Ch. 17). The growing shortfall in energy needs must be narrowed by cutback in global economic activity, by stringent energy efficiency measures or by alternative non-fossil fuels.

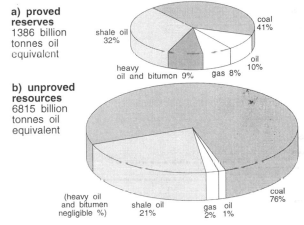

Figure 12.3 Estimated world fossil fuel resources.

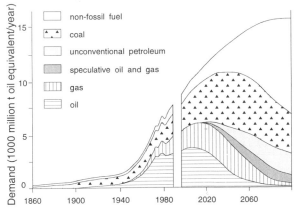

Figure 12.4 An estimate of future world energy demand and the fossil fuel shortfall.

12.3 Uranium and nuclear power

Uranium is a special kind of geological fuel, releasing energy by radioactive decay rather than by combustion. This decay process is very potent, with 1 kg of the isotope ^{235}U releasing three million times the energy in 1 kg of coal. Consequently uranium was seen during the 1950s and 1960s as the ideal fuel of the future to replace shrinking reserves of conventional fossil fuels. More recently concern has grown about the safety of all parts of the nuclear fuel cycle (Fig. 12.5), particularly at the power stations and at the sites of nuclear waste storage and disposal. The rapid growth of nuclear power in the 1970s and 1980s has slowed until safe solutions can be found at an economic cost.

Most nuclear reactors harness the heat energy released when ^{235}U breaks down under neutron bombardment. Each ^{235}U atom captures one neutron to form the highly unstable ^{236}U, which quickly decays to fission products such as yttrium and iodine, themselves radioactive. Further neutrons are released to bombard more ^{235}U atoms. This chain reaction would give a nuclear explosion if some of the neutrons were not absorbed by neutron **moderators** such as graphite or heavy water placed between the uranium fuel elements in the reactor **core**.

The heat energy from fission is used to convert water into steam for driving electricity-generating turbines. In a **pressurized water reactor** (PWR) this same water is used to cool the reactor core. In a **magnox** reactor and an **advanced gas-cooled reactor** (AGR), there is an intermediate cooling circuit using carbon dioxide gas. Britain has nine magnox reactors and seven AGRs for public power generation, with one PWR being built (Fig. 12.11).

Natural uranium contains only 0.7% ^{235}U and is mostly composed of the more stable ^{238}U. But this isotope can also be utilized by bombarding it with the high-energy **fast neutrons** absorbed by the moderator in normal reactors. The ^{238}U is converted to ^{239}U which decays through ^{239}Pu to products such as zirconium and xenon. Surplus plutonium is produced, so that reactors using fast neutrons are called **fast breeder reactors**. They theoretically produce about a hundred times more energy from uranium than a conventional **thermal reactor**, but they present formidable engineering problems which are yet to be overcome for commercial use.

The nuclear reactor is the key link in a much larger nuclear cycle (Fig. 12.5). The uranium ore, containing between 0.025% and 0.25% uranium, is milled to a fine sand. The uranium is extracted by solvents to form **yellowcake** with about 85% uranium. The yellowcake is chemically processed to yield either uranium metal, for magnox reactors, or uranium hexafluoride (**hex**), centrifugally enriched to 2.3% ^{235}U for other burner reactors. After use in the reactor core the spent fuel elements are reprocessed to extract unused uranium and plutonium for reuse or strategic storage. Most parts of the cycle, but particularly reprocessing, produce radioactive wastes.

Geologists are concerned with three stages of the nuclear cycle (Fig. 12.5): supplying the raw uranium (this section), monitoring the geohazard risk to nuclear installations (Ch. 4), and assessing the safety of underground sites for waste repositories (Ch. 16).

Economic natural deposits of uranium are produced by a range of geological processes (Fig. 12.6). These must concentrate uranium by at least a hundred times its 3 ppm average abundance in the upper continental crust. So-called **primary uranium deposits** are concentrated by internal earth processes and **secondary uranium deposits** by surface processes which operate on eroded primary material. However, high-U rocks have probably passed through several cycles where "secondary" deposits are reburied and partially melted to source new "primary" deposits.

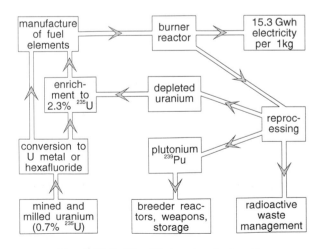

Figure 12.5 Simplified nuclear fuel cycle.

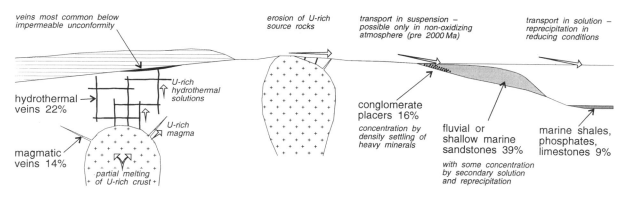

Figure 12.6 Formation processes of uranium ore deposits with percentage of the world reserve formed by each process.

Magmatic veins form from residual magmatic liquids, in which uranium is concentrated because its ionic size and charge make it incompatible with common rock-forming minerals. The uranium occurs as the oxide **uraninite** (U_3O_8) or substitutes for the calcium in apatite and the zirconium in zircon. **Hydrothermal veins** contain massive uraninite called **pitchblende**, deposited from hot magmatic water flushing through joints and fault zones. These fluids are commonly concentrated below impermeable unconformities, forming lenticular ore-bodies in structural traps.

Uplift and erosion of primary uranium deposits releases uraninite for surface transport. However, this mineral breaks down quickly in an oxidizing environment and the uranium goes into solution. Only in the low-oxygen atmosphere of the early Earth was transport and sedimentation of clastic uraninite possible, resulting in **conglomerate placers**. These often contain other heavy minerals such as gold and pyrite. In more recent oxidizing atmospheres (post 2000 Ma) uranium has been transported only in solution. It is precipitated again in reducing conditions in ground waters or surface waters. For this reason uranium is particularly associated with black shales, coals and phosphates. Rich uranium ores also occur at oxidation fronts in permeable sandstones following solution and reprecipitation.

Britain has only low-grade uranium ores which are at present uneconomic. The main deposits are hydrothermal veins associated with mineralization in southwest England and southern Scotland, and phosphatic, carbonaceous and fault breccia ores in the Devonian of the Scottish Highlands.

World reserves and production are apparently concentrated in a few countries (Fig. 12.7), predominantly in Precambrian cratons. However, political as well as geological factors contribute to this, with the strategic importance of uranium resulting in unreliable statistics for China, the former Soviet Union and eastern Europe. At present levels of consumption and in current economic conditions, proved reserves of uranium would last about 50 years. This lifetime will surely be lengthened by new ore discoveries, but shortened by any renewed global expansion of nuclear power generation. These data emphasize that uranium is also a finite resource, with only the commercial development of fast breeder reactors able to extend its lifetime much beyond that of oil and gas.

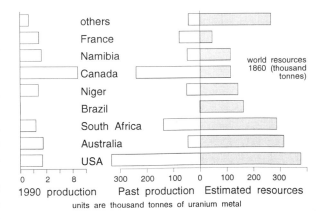

Figure 12.7 World uranium production and estimates of low to medium-cost "reasonably assured resources". No data available from China or former Soviet Union.

a) Hyperthermal system

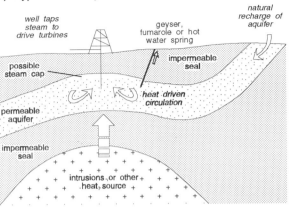

b) Geothermal aquifer

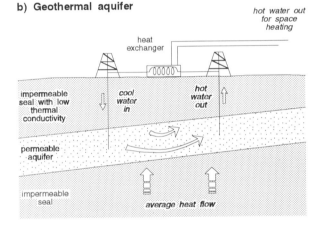

c) Geothermal hot dry rock

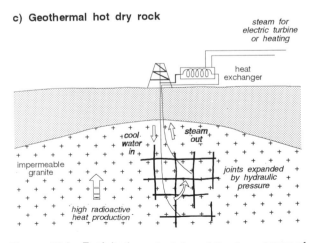

Figure 12.8 Exploitation principles of the three types of geothermal energy.

12.4 Geothermal energy

Heat energy flows to the Earth's surface from two sources: from the decay of long-lived radioactive isotopes concentrated in the upper crust, and from the residual heat of the lower crust, mantle and core. On average, the surface heat flow is about $60\,\mathrm{mW\,m^{-2}}$ (milliwatts per square metre). Even in such an average area of crust, this heat flow could be exploited for low-temperature applications such as space-heating. But geothermal energy is economically more attractive in areas of above-average thermal gradient, where high temperatures are reached at shallow depths. Three different types of geothermal system are exploited, differing in their subsurface temperature and the nature of their host rocks.

Hyperthermal systems, or **high enthalpy systems**, have very high heat flows, around $300\,\mathrm{mW\,m^{-2}}$, owing to hot magmas actively intruding high in the crust (Fig. 12.8a). Groundwater is heated to boiling point or above and moved around by **hydrothermal convection**. Fracture systems allow hot water or steam to escape to the surface as **hot springs**, **geysers** or **fumaroles**. For power production this groundwater must be trapped in a way analogous to oil and gas: a permeable **aquifer** overlain by an impermeable **seal** and arranged in a geometrical **trap** such as an anticline (Fig. 12.8a). The aquifer is drilled to tap the water, at up to 250°C and 40 atmospheres pressure. A **steam cap** may be present in the trap and even if not the superheated water "flashes" to steam as it ascends to lower near-surface pressures. This steam is piped to drive turbines for electricity generation.

In many hyperthermal systems the natural heat flow is so great that the resource is effectively renewable on a human time scale. The rate of energy extraction is more likely to be constrained by the rate at which groundwater is replenished by surface precipitation at its **recharge zone** (Fig. 12.8a). Hyperthermal systems are restricted to regions with a plate boundary or intraplate volcanism, power plants being concentrated in North America, the Phillipines, Italy, Japan and New Zealand. These plants currently account for only 0.1% of world electricity production, but this figure is rising rapidly. Geothermal energy prospects in the British Isles must rely on other types of system.

Geothermal aquifers, or **low enthalpy systems**, occur in sedimentary sequences with near average heat flow but with a higher than normal thermal gradient. This occurs if the sequence contains rocks with low **thermal conductivity**, typically mudrocks, which act as insulation to deep-sourced heat flow (Fig. 12.8b). Any permeable aquifers in the sequence then contain hot groundwater, ideally above 60°C, at drillable depths of only a couple of kilometres. The resource is exploited by pumping up the hot saline water, and passing it through a heat exchanger. The secondary water circuit feeds low-temperature applications such as space heating. The cooled saline waters are reinjected to the host aquifer about a kilometre away, maintaining the water pressure in the aquifer.

The heat flow in low enthalpy systems is insufficient to maintain an economic temperature in the aquifer for more that a few tens of years. The thermal resource is therefore non-renewable, and its exploitation amounts to **heat mining**. However, such resources are globally widespread. Potentially economic systems in Britain occur in thick post-Carboniferous sequences with suitable aquifers (Fig. 12.9). Experimental schemes have been operating in Southampton, Northern Ireland and Lincolnshire. The identified UK resources from the Permo-Triassic aquifer alone equate to about 3500 million tonnes of oil, about 16 times the annual UK energy demand. Much of this resource is currently subeconomic.

Geothermal hot dry rock systems exploit the temperatures of 150°C–200°C that occur at depths of around 5 km in rocks with higher heat production than average. Typically these are granites containing high concentrations of radioactive uranium, thorium and potassium. Their heat energy can be extracted by a paired boreholes arrangement deviated from a single wellhead (Fig. 12.8c). However, granites have a low natural permeability and groundwater content. To yield economic amounts of energy, the permeability must be enhanced by injecting high-pressure water or using explosives to widen the natural joint systems between the two holes. In a working system the extracted steam could drive electric turbines through an intermediate heat exchanger.

Three areas in the British Isles have been identified as hot dry rock prospects. Most suitable are the Variscan granites of southwest England, then the Caledo-

subsurface extent of high heat flow granites (heat production generally >4 mW m⁻³)

subsurface extent of potential geothermal aquifers (mean temperatures >40°C and transmissivity >10 darcy metres)

Figure 12.9 Potential geothermal fields in the UK.

nian granites of northern England, and less important the Caledonian granites of the eastern Scottish Highlands (Fig. 12.9). An experimental project using 2 km deep boreholes has been run in Cornwall. It has demonstrated the feasibility of the strategy but also the technical difficulty of ensuring adequate permeability through and only through the high temperature part of the rock body.

Despite these problems and its ultimately non-renewable nature, the hot dry rock resource is likely to prove important in the future. The total accessible UK resource base, defined as the heat stored at less than 7 km and at more than 100°C, is estimated at 36×10^{21} joules. Even if only 1% of this energy is recoverable, this equates to about 8 billion tonnes of oil, around 40 times the annual primary energy consumption of the UK.

12.5 Renewable energy

The energy from conventional nuclear reactors and from most geothermal sites is, like fossil fuels, finite on a human timescale. To meet growing world energy needs, society must look to forms of energy that are truly renewable. Whereas geologists are involved in several parts of the nuclear and geothermal cycles, they play a more marginal role in developing renewable energy schemes, for instance in assessing their engineering problems and environmental impacts. However, an account of renewables here will allow their role in future energy strategy to be assessed.

Hydroelectricity was the first renewable energy source to be exploited in the British Isles (Fig. 12.10). It supplies annually about 2.5% of primary energy (0.2 million tonnes oil equivalent, mtoe) in the Irish Republic and about 0.7% (1.5 mtoe) in the UK, mostly in Scotland (Fig. 12.11). Most sites suitable for large hydroelectric schemes have already been developed, and future potential lies with small-scale installations that utilize heads of water as low as 3 metres. These will add only another 0.5 mtoe per year to the UK energy supply, most of it again from Scotland.

Wave energy is abundant around the British Isles (Fig. 12.11). It could supply an estimated 6% of UK primary energy (12.5 mtoe, Fig. 12.10) and as much as four times the demand in Ireland. However by its nature the resource is diffuse, and requires long arrays of collectors to tap efficiently. Such a strategy is feasible but expensive offshore. Generators to collect shoreline wave energy are more practical but exploit only a fraction of the available resource.

Solar energy potential is considerable in the British Isles even though the mean daily duration of bright sunshine varies from five hours, in coastal areas of southern England, down to less than three hours, in the Scottish Highlands. The estimated available resource in the UK is about 6% of its energy use (12.8 mtoe, Fig. 12.10). Nearly half this total could be achieved by better design of buildings to give a passive heat gain. This strategy is currently more attractive economically than either active solar heating, using heat-absorbing panels, or electricity generation, using photovoltaic cells.

Tidal energy can be exploited by building barrages across major estuaries. The resulting head of water is used to drive turbines, both on the ebb and the flow of

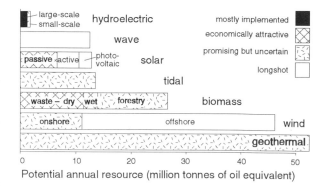

Figure 12.10 The potential for alternative energy in the UK.

the tide. The most suitable estuaries in the UK (Fig. 12.11) could provide as much as 6.5% of present national energy requirements (13.5 mtoe, Fig. 12.10). These schemes are likely to become economically viable as fossil fuel prices rise.

Biomass of various types can be used for energy generation. Purpose-grown forestry crops are currently a less economic fuel than waste materials from industrial, agricultural and domestic sources. Dry wastes are burned directly, whereas wet wastes are bacterially decomposed to yield methane gas for combustion. The yield of methane gas from UK landfill sites is already 0.1 mtoe, but the national biomass potential is about 27 mtoe, about 13% of UK energy use (Fig. 12.10).

Wind energy is the most abundant renewable resource, particularly in the north and west of the British Isles (Fig. 12.11). The UK potential is about 22% of its energy use (46 mtoe, Fig. 12.10), although only a quarter of this could be exploited by onland rather than offshore wind turbines. Ireland has an even greater wind potential, up to five times its national energy demand.

Any one of the renewable energy strategies can provide only a percentage of present national energy needs. However, the total renewable energy resource in the UK is about 113 mtoe, over half the primary energy need. The geothermal potential raises the contribution by a further 53 mtoe, a substantial alternative to the present dependence on fossil fuels. Schemes operating or being constructed in the early 1990s will exploit about 2.7 mtoe of this potential.

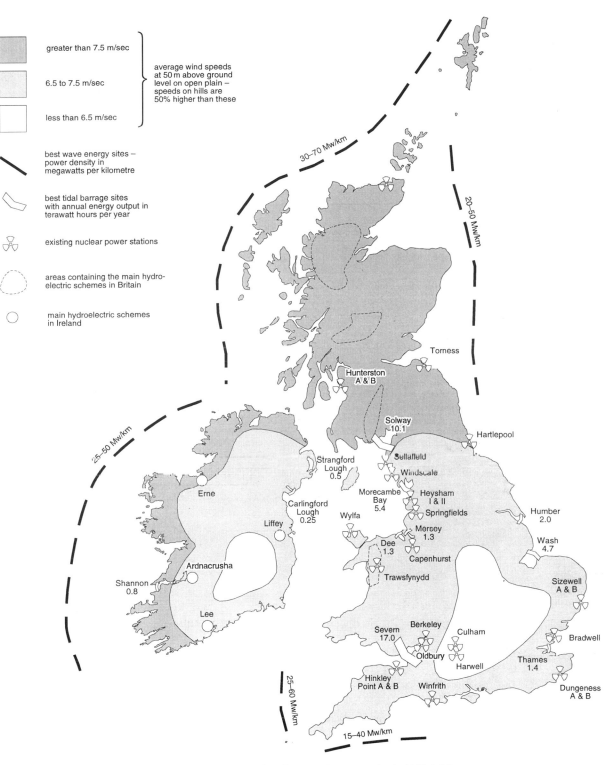

Figure 12.11 Geography of alternative energy in the British Isles.

101

12.6 Energy futures

A global perspective is essential in predicting future trends in energy use in the British Isles. Although the region is relatively rich in fossil fuels, their price and profitability are strongly controlled by the international nature of the energy market. All views of the next century (e.g. Fig. 12.4) predict a growing inadequacy of fossil fuels to satisfy increasing world energy demands. This global shortfall of oil, gas and coal will raise the price of domestic supplies and make alternative energy strategies increasingly profitable.

Even this international view is inadequate, reflecting the preoccupation with shortage of resources which has dominated energy forecasts this century. In the next century the major constraint on the use of fossil fuels is more likely to be the shortage of an adequate atmospheric sink for their gaseous effluents; carbon dioxide, nitrogen oxides, sulphur dioxide, methane, particulate matter and volatile organic compounds. The impacts of fossil fuel use are detailed later (Ch. 17), but their implications for future energy trends are simple. There will be strong social and economic pressure to use more of the clean renewable energy strategies at the expense of fossil fuels.

One feasible plan for the UK shows the kind of energy future that must be adopted by industrialized nations to allow the rest of the world to develop without catastrophically overloading the global atmosphere (Fig. 12.12). Increasing use of the large renewable resource is one key factor in this plan. The other is a 40% cut in total energy use achieved by greater restraint and efficiency.

Further reading

Allen, J. E. 1992. *Energy resources for a changing world*. Cambridge: Cambridge Univeristy Press. [A straightforward introduction to alternative and conventional energy sources.]

Blunden, J. & A. Reddish 1991. *Energy, resources and environment*. Sevenoaks, Kent: Hodder & Stoughton/Open University. [Contains excellent chapters on energy resources, impacts and futures.]

British Petroleum 1992. BP *statistical review of world energy*. London: British Petroleum. [Annual compilation of production and consumption data for fossil fuels, nuclear energy and hydroelectricity.]

Brown, G. C. & E. Skipsey 1986. *Energy resources: geology, supply and demand*. Milton Keynes, Bucks.: Open University Press. [Part II covers alternative energy resources.]

Cartledge, B. (ed.) 1993. *Energy and the environment*. Oxford: Oxford University Press. [Collected lectures offering a complete spectrum of views on the future of energy supply.]

Cassedy, E. S. & P. Z. Grossman 1990. *Introduction to energy: resources, technology and society*. Cambridge: Cambridge University Press. [Quantitative yet readable coverage of the principles and methods of energy generation.]

Central Office of Information 1992. *Energy and natural resources*. London: HMSO. [The UK government view of its energy supply strategy.]

Energy Technology Support Unit 1989. *Renewable energy in Britain*. London: Department of Energy [Regularly updated summary of renewable energy prospects in the UK.]

Downing, R. A. & D. A. Gray 1986. *Geothermal energy - the potential in the United Kingdom*. London: Her Majesty's Stationery Office. [The British Geological Survey assessment of geothermal potential.]

Foley, C. 1992. *The energy question,* 4th edn. Harmondsworth: Penguin. [An informed review of energy options from a technical, environmental and political viewpoint.]

Patterson, W. C. 1986. *Nuclear power,* 2nd edn. Harmondsworth: Penguin. [An informative reference, including a comprehensive bibliography.]

Patterson, W. C. 1990. *The energy alternative: changing the way the world works*. London: Boxtree. [A popular, informed coverage of alternative energy prospects.]

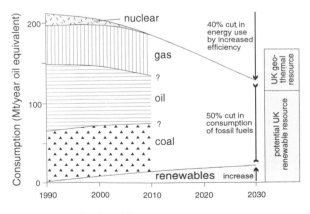

Figure 12.12 An achievable energy strategy for the UK.

IV HUMAN IMPACTS ON THE EARTH

The potential for humans to harm their environment has, for most of history, been thought negligible by comparison with that of natural processes. This complacent view is changing fast, in the face of wide-spread local impacts and threatened global effects. This part describes those impacts which involve geology (Ch. 13), either because they result from the production or use of geological resources.

Chapter 14 deals with the effects of **resource extraction** – mining, quarrying and production of petroleum. Chapter 15 considers the many ways of engineering the geological environment – building, tunnelling, agriculture, and altering slopes, coasts and rivers. The part ends with two chapters on the wastes from society. Those mostly solid and liquid wastes that are disposed of on or under the surface are covered in Chapter 16. Chapter 17 describes the effects of the mainly gaseous wastes emitted into the atmosphere. The threat of global warming from the enhanced greenhouse effect is the most serious of impacts from the use of geological resources.

The cooling towers of a coal-fired power station at Ferrybridge, Yorkshire. Such plants produce most of the sulphur dioxide in Britain, together with nitrogen oxides and smoke.

Chapter Thirteen

ENVIRONMENTAL IMPACTS

13.1 The nature of the environment

Previous parts of this book treated the environment as a control on human settlement and activity (Pt II) or as a passive resource for human exploitation (Pt III). They largely ignored the changes to the environment that human activity causes. The next four chapters deal with these human impacts, concentrating on those in which geological processes or resources play some part.

But what is "the environment" and what role has geology in understanding its workings? This chapter looks at these questions from several general perspectives, as a background to the detailed chapters that follow. Relevant issues of environmental ethics and philosophy are deferred until Part V.

The Earth is conventionally subdivided into a series of crudely concentric zones (Fig. 13.1). The **lithosphere**, **hydrosphere**, **biosphere** and **atmosphere** are dominantly composed of rock, water, organisms and gases respectively. The **ecosphere** comprises the biosphere and those parts of other zones with which organisms interact. The ecosphere essentially corresponds to the **environment** as it is most widely understood. It includes the lower parts of the atmosphere, where some animals live and where the gaseous residues from organic activity are reprocessed. It encompasses virtually all the hydrosphere – the oceans, seas, rivers and lakes. Most relevant to geologists, it includes the upper part of the lithosphere – the veneer of soil and marine sediment in which organisms live and die, and the subsurface rocks at least as far downwards as groundwater can percolate.

The workings of the present-day environment are subdivided into a range of biological, physical and chemical systems, each the speciality of a distinct group of scientists such as zoologists, botanists, hydrologists, oceanographers and climatologists. The **geological environment** of the present-day Earth is regarded as the subsurface solid rocks and their included water or petroleum fluids (Fig. 13.1). The surface veneer of decomposed rocks is a shared responsibility with soil scientists.

The "geological environment" extends far below the ecosphere, right to the Earth's core. These deeper zones are below the range of biological and human impacts, but they play host to important environmental factors – deep earthquakes in the lithosphere, volcanic activity originating in the underlying **asthenosphere**, and the magnetic field generated in the convecting outer **core**.

In Earth history, the term "geological environments" assumes an even wider meaning, inclusive of surface as well as subsurface systems (Fig. 13.1). Ancient rocks preserve information, not just about conditions in the lithosphere, but about seawater chemistry, biodiversity, atmospheric composition and many other environmental factors. Geologists are the environmental scientists of the old Earth. This knowledge of ancient global change can be used to illuminate present environmental impacts and their development into the future. This predictive use of Earth science makes it difficult to draw a firm line around geologically relevant impacts.

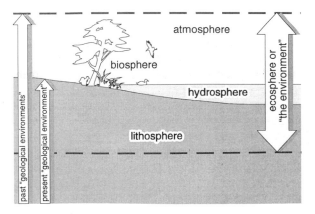

Figure 13.1 Different concepts of the environment.

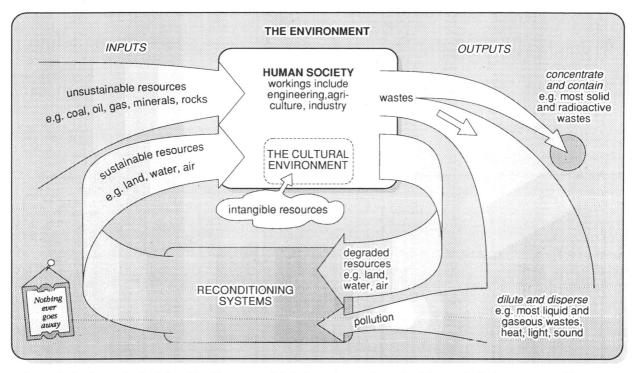

Figure 13.2 Conceptual relationship of human society to its environment, emphasizing geological resources and impacts.

One way of assessing human impacts is to regard society as a separate **subsystem** within the wider environment **system** (Fig. 13.2). Matter and energy fuel the **workings** of the human subsystem as **inputs**, and are returned to the environment as **outputs** of waste and used resources. Each part of this cycle perturbs the natural environment.

The inputs to society are of two main kinds (Ch. 5.1). **Unsustainable resources** cannot be replenished by natural processes within a human timescale. Geological examples include mineral deposits and fossil fuels. **Sustainable resources**, such as land and water, can be reconditioned by the environment, potentially as fast as they are used. The extraction of all these resources has environmental impacts (Ch. 14).

The workings of human society – activities such as engineering, agriculture and industry – create their own impacts (Ch. 15). Indeed, the purpose of much of this activity is specifically to fashion the environment in ways seen as beneficial to society – to create agricultural land, structures for shelter, transport systems for communication. However, many of these intended enhancements bring accompanying and often unfore-

seen degradation. If nothing else, they affect **intangible resources** such as scenery and biological diversity, assets for which it is difficult to give a monetary value (Fig. 13.2). These resources add to the **cultural environment**, which is such a distinctive feature of human society.

The output side of the human cycle includes two main components (Fig. 13.2). **Degraded resources** such as water, land and air require reconditioning by the environmental system, ready for re-input to society. **Wastes** are society's unwanted matter and energy and are disposed of in two contrasting ways. Those wastes thought too toxic for release to the environment are concentrated and contained by supposedly leak-proof barriers. Nuclear wastes and some chemicals are examples (Ch. 16). Most wastes, however, are injected into environmental systems expected to dilute and disperse them, particularly the hydrosphere (Ch. 16) and the atmosphere (Ch. 17). The capacity of the environment to handle such wastes is often overestimated, leading to **pollution** which can threaten the vital environmental reconditioning systems on which society depends (Fig. 13.2).

13.2 Geological impacts

The simplified cycle of human interaction with the environment (Fig. 13.2) provides a framework for describing geologically related impacts – first in the provision of inputs (Ch. 14), then during the working of society (Ch. 15), and lastly from the effects of human outputs to the Earth surface or subsurface (Ch. 16) and to the atmosphere (Ch. 17). However, this scheme ignores the complexity of most environmental impacts. The exploitation of petroleum provides a geologically related example of this complexity (Fig. 13.3).

Outputs of waste and degraded resources occur at all stages in this industrial process from its **upstream** end of extraction to its **downstream** end of final uses. Some of these petroleum impacts affect the geological environment itself. Ground subsides where oil and gas are pumped out from weak rocks. Oil leaks into surface or groundwater from drilling operations, tanker spills, and broken pipelines or storage tanks. Water is also polluted by the products of petroleum, either directly from petrochemical plants or from byproducts

such as fertilizers, pesticides and detergents. Soils receive deposits of soot and nitrous oxides. Land and seafloor are degraded by ports, refineries, petrochemical plants and pipelines, and by the power stations, roads and airports that form the downstream infrastructure of petroleum use.

The atmospheric impacts are even more varied and dangerous. Methane escapes during drilling or from leaking pipelines. Other volatile hydrocarbons are emitted from refineries, petrochemical plants, power stations and petroleum-fuelled vehicles. Halogenated hydrocarbons such as CFCs (chlorofluorocarbons) are given off during their manufacture and use. Sulphur dioxide, nitrogen oxides and, most seriously, carbon dioxide are all emitted when petroleum is burned (Ch. 17).

Other impacts affect the internal quality of society as much as its external environment. Lives are lost in accidents on oil rigs, during industrial processing, and during use, particularly on the roads. Noise pollution, the design of cities, international relations – these are less direct but no less pervasive impacts of "the hydrocarbon culture" of the 20th century.

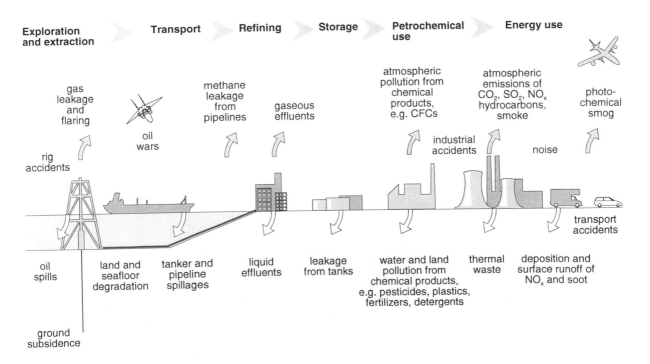

Figure 13.3 Some of the environmental impacts of exploiting oil and gas.

The petroleum cycle emphasizes that few environmental impacts are uniquely geological. In particular, regional or global effects usually involve **dispersion** of local pollutants by the atmosphere or oceans (Fig. 13.4). Some geological processes, such as landslides, earthquakes and volcanic flows, are suitably fast but are local in nature. More widespread processes, such as plate motion or heat flow, operate too slowly to act as dispersants on a human timescale. Groundwater flow is locally effective, but rivers, ocean currents and winds are the agents of global pollution (Fig. 13.4). Migrations of animals and human beings or the diseases that they carry are also a major factor in the global environmental balance. Fastest and most potent of all are radio and television, promoting global awareness of more local environmental problems.

Environmental impacts themselves form a spectrum in time and space (Fig. 13.5). Within this spectrum, the localized, short-term effects, such as air and surface water pollution, provoke the most public concern and action in a wealthy region such as the British Isles. Subsurface "geological" impacts, such as groundwater pollution, degraded land and nuclear waste disposal, pose local threats over many years or even centuries, and raise issues of **intergenerational equity** – responsibility to future inhabitants of the region. Global impacts such as climate change, induced by the waste gases from fossil fuels, must be assessed in a framework of **inter-regional equity**. Chapter 18 deals further with these two ethical aspects of environmental analysis.

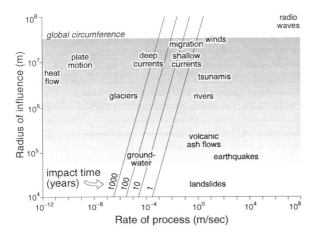

Figure 13.4 The rates and spatial influence of some environmental dispersal processes.

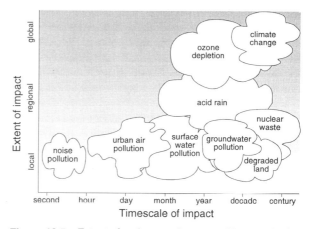

Figure 13.5 Extent of various environmental impacts in time and space.

Further reading

Department of the Environment 1992. *The UK environment.* London: HMSO. [A useful UK government view of environmental priorities in Britain.]

Goudie, A. 1990. *Human impact on the natural environment,* 3rd edn. Oxford: Basil Blackwell. [A comprehensive overview of environmental impacts worldwide.]

Keller, E. A. 1992. *Environmental geology,* 6th edn. New York: Macmillan. [Contains much thoughtful material on the nature of geological impacts to the environment.]

Lumsden, G. I. (ed.) 1992. *Geology and the environment in Western Europe.* Oxford: Oxford University Press. [Chapter 5 is a helpful catalogue of geologically related environmental impacts.]

Newson, M. (ed.) 1992. *Managing the human impact on the natural environment.* London: Belhaven. [A collection of articles providing an overview of environmental impacts.]

Pearce, D., A. Markandya, E.B. Barbier 1989. *Blueprint for a green economy.* London: Earthscan. [A readable introduction to environmental economics, including the problem of valuing the environment.]

Revelle, P. & C. Revelle 1988. *The environment: issues and choices for society.* Boston: Jones & Bartlett. [One of many well illustrated introductions to environmental science and human impacts.]

White, I. D., D. N. Mottershead, S. J. Harrison 1992. *Environmental systems: an introductory text,* 2nd edn. London: Chapman & Hall. [Excellent introduction to environmental processes in a framework of systems analysis.]

Chapter Fourteen

RESOURCE EXTRACTION

14.1 Extracting geological resources

The appropriate method of extracting a subsurface resource depends on its depth of burial and its physical state – whether it is a fluid, a consolidated rock or an unconsolidated sediment (Fig. 14.1). Fluids such as water, oil and gas are pumped out through **wells** drilled into the permeable reservoir rocks that contain them. Deeply buried rocks are reached by **underground mines** and are brought to the surface through interconnected tunnels and shafts. In the British Isles, most coal and some metallic and industrial minerals are won in this way. However, it is often cheaper to extract these commodities by **surface mining** of shallow deposits. **Quarrying** refers to surface mining of hard rocks for use as construction materials. Un-

consolidated sediments are exploited by other **surface extraction** methods such as digging of peat and dredging of sand and gravel.

Each extraction technique has its particular environmental impacts (Fig. 14.1). This chapter will concentrate on **geomorphological** effects, those that alter the shape and height of the land surface. These effects include the depressions due to surface excavation or subsidence over deep mines, and the mounds and hills formed of solid wastes. Toxic chemical effects of such wastes and of the liquid and gaseous wastes produced by some extraction methods are covered in Chapters 16 and 17. Those chapters also describe some of the effluents from **beneficiation** of resources – processes such as the leaching or smelting of metallic ores.

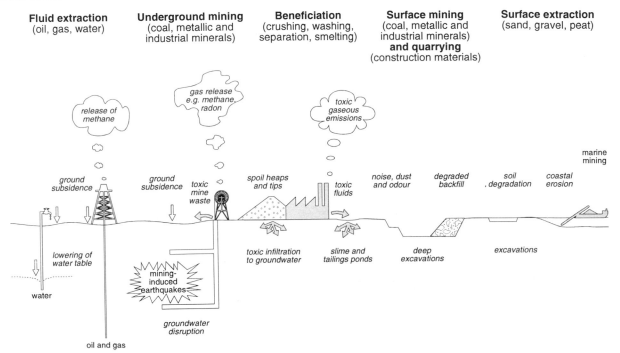

Figure 14.1 Methods of extracting geological resources together with their main environmental impacts.

14.2 Withdrawal of subsurface fluids

Water, oil and gas resources occur within the pores or fractures in rocks. When these water **aquifers** or petroleum **reservoirs** are penetrated by a well, the pressure of the fluid may be enough to force it to the surface, or this pressure may need augmenting by pumping (see Ch. 7 for water and Ch. 11 for oil and gas). The most direct effect of extraction is progressively to lower the natural pressure, the **piezometric head**, in the fluid itself.

The Chalk aquifer below London suffered such a pressure drop (Fig. 14.2). Until the middle of the nineteenth century, recharge of this aquifer at its outcrop around the edge of the London Basin was sufficient to maintain near-artesian flows in its centre – that is, the water level in wells was close to the ground surface (Fig. 14.2b). Recharge could not keep pace with the rapid increase in abstraction during the later nineteenth and early twentieth century (Fig. 14.2a). The piezometric head dropped by as much as 70 m (Fig. 14.2c), rendering older wells ineffective without increased pumping.

A second effect of fluid withdrawal is enhanced compaction of the aquifer or reservoir and associated sediments, particularly clays. Lowered pore fluid pressures cause the grain framework to bear more of the weight of overlying rocks. In a typical sandstone or limestone aquifer, compaction is restricted to small elastic deformation of the grains. Adjacent mudstones, subject to the same pore pressure decrease, suffer greater compaction by sliding and rotation of the grains, packing them more closely and decreasing the rock volume.

The most serious impact of compaction is subsidence of the ground surface. In the offshore Ekofisk oilfield (North Sea; Fig. 11.12b) the sea bed sank over 2.5 m between 1971 and 1985 by compaction of the weak Chalk reservoir, an amount that would have had serious consequences at an onshore site. Since 1985, subsidence has been reduced by injection first of gas and then of water. The relatively strong clay rocks beneath London restricted maximum ground subsidence there to about 25 cm. More damaging to tunnels and foundations has been the re-swelling of the clays due to rising water levels, as abstraction from the Chalk has slowed since 1970.

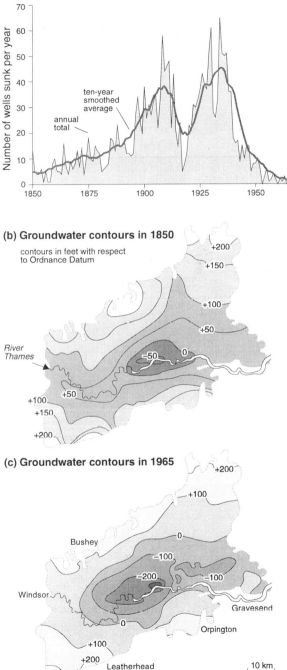

(a) Annual construction of wells into the Chalk aquifer

(b) Groundwater contours in 1850

(c) Groundwater contours in 1965

Figure 14.2 Wells sunk into the Chalk aquifer below London in the period 1850–1965 and the resulting lowering of the piezometric surface.

14.3 Underground mining

Longwall mining is the main method of extracting coal in Britain (Fig. 14.3a); it can be used for any laterally continuous rock body with a uniform thickness and a gentle dip. Coal is removed by a track-mounted cutter moving along a face several hundred metres long. The cutter operates beneath a roof supported by hydraulic jacks, which are slid forwards after the cutter has passed. The roof behind the jacks collapses onto the former floor of the coal seam, the **goaf**. The face can advance by up to a kilometre each year. Access to it is maintained by tunnels joining each end of the face to the mine haulage roadways.

The roof collapse behind a longwall face propagates upwards and outwards through the overlying rock with a geometry measured by the **angle of draw** (Fig. 14.3a). This varies with the rock strength but is roughly 30°, resulting in a **subsidence bowl** at the ground surface considerably wider than the **extracted panel** of coal. The maximum depth of the subsidence bowl is always less than the seam thickness, because

of the volume increase as cracks open up within the subsiding rocks. Also damaging to built structures is the ground tilt as the subsidence wave passes, and the related cycle of surface extension and shortening. However, these effects were more severe with older shallow mining than during modern mining of deep seams. Moreover, the pattern and timing of subsidence over longwall faces is predictable, so that structures at risk can be strengthened before mining begins.

Pillar-and-stall working is also suited to gently dipping beds (Fig. 14.3b). The deposit is only partially removed, leaving intervening pillars to support the roof. The pillars are elongate or square, and are spaced to allow extraction of between 50% and 85% of the bed. In North American coal mines the pillars are removed on retreat from the seam, allowing roof collapse similar to that of a longwall face. However, pillars have been left in place in most mines in the British Isles. These include modern gypsum mines, and old mines for coal, building stone, ironstone and clays. The old mines present a serious hazard to development

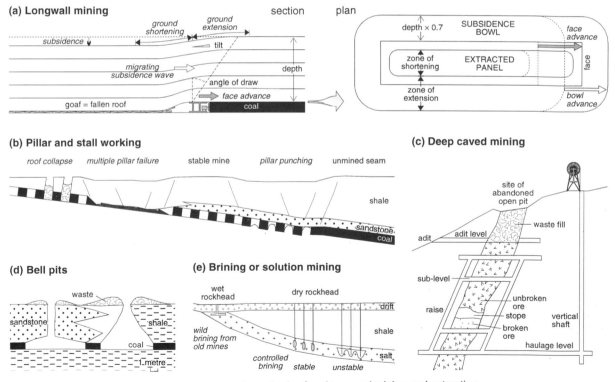

Figure 14.3 The main methods of underground mining and extraction.

in old mining areas. An example from Suffolk (Fig. 14.4) involved **roof collapse** into the passages of underlying chalk mines at 10–12 m depth. Collapse was triggered by changes in groundwater flow induced by building development in the late 1960s. Other types of failure can include **multiple pillar failure**, and **pillar punching** into weak roof or floor rocks (Fig. 14.3b).

Bell pits are a yet older form of mining, mostly dating from before 1700 (Fig. 14.3d). Shafts were sunk to extract unsupported circular areas in an underlying seam, the shape of the pit depending on the strength of the roof rock. Bell pits are rarely more than 10 m deep, so that they represent a localized and distinctive subsidence hazard.

Deep caved mining is used to extract steeply dipping and irregular mineral deposits, typically metallic ores. One common method is to mine the ore body from below, through a vertical **shaft** and horizontal **levels** (Fig. 14.3c). The first ore is removed to form a void or **stope**. Later ore is either allowed to remain as a working floor to the stope, eventually to be tapped off from below, or is replaced by **waste fill** from the surface processing plant. Shallow ore can be accessed without a shaft through an **adit**, and mined upwards or downwards. Any deep caved technique may produce a subsidence bowl at the ground surface as abandoned stopes collapse by caving of their roofs. Only thorough backfilling can prevent this, a rare practice in old mines unless necessary for the mining strategy.

Brining is the pumping of dissolved salt from underground evaporite beds (Fig. 14.3e). Extracted water is replaced in **wild brining** by natural groundwater and in **controlled brining** by injected freshwater. Controlled brining can produce stable cavities that cause ground subsidence only if allowed to coalesce. Wild brining is less predictable and has produced large subsidence zones in the Cheshire saltfield, often elongated over subsurface water streams (Fig. 14.5). Even more damaging methods, before they were banned about 1930, were pillar-and-stall mining with excessive extraction ratios and **bastard brining** – pumping of water from the abandoned mines. Sinkholes about 100 m wide and 10 m deep formed catastrophically as remaining pillars dissolved and collapsed, causing major property damage.

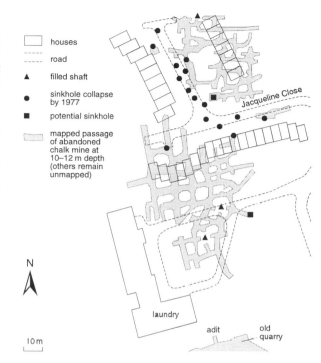

Figure 14.4   Old chalk mines below Bury St Edmunds.

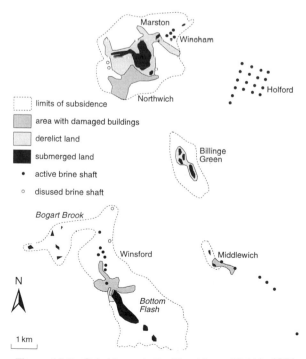

Figure 14.5   Subsidence in the Cheshire saltfield in 1954.

14.4 Surface mining and quarrying

Surface mineral working is now the most disruptive form of extraction in the British Isles, increasing at least fivefold during this century, while underground mining activity has halved. Surface workings disturb all the land underlain by the economic deposit, whereas subsurface mining can be carried out beneath land used for other purposes.

Some terms for describing surface workings relate to the nature of the deposit. A **mine** exploits minerals or fossil fuels, a **pit** poorly consolidated rocks such as clay, sand or coal, and a **quarry** hard rocks such as sandstone, limestone and slate. More relevant to environmental impacts is the method of surface working (Fig. 14.6). This depends on the depth and orientation of the deposit and on the volume ratio of the usable resource to waste material.

Area strip mining exploits gently dipping beds within tens of metres of the ground surface (Fig. 14.6a). Carboniferous coals (Ch. 10) and Jurassic ironstones (Ch. 9) in particular have been mined in this way in Britain. Strip mining starts with an elongate trench along the outcrop of the productive bed. To exploit its subsurface continuation, the unproductive **overburden** is then stripped off and dumped on the ground already mined. The working trench advances across strike, leaving in its wake a **hill-and-dale** land-

scape of waste ridges. Before 1951, there was no statutory obligation in the UK to reclaim such land, but now it must be naturally contoured, covered with topsoil, and revegetated. This controlled strip mining produces total localized short-term disruption, but at least leaves land suitable for agriculture or building. A positive effect of some strip mining for coal in Britain has been the restoration of derelict land left by former mining activity.

A form of strip mining rarely practised in the British Isles is **dredging** for **placer** deposits, particularly of tin or gold, in alluvial sediments.

Contour strip mining has been used where gently dipping deposits occur in hilly areas (Fig. 14.6b). In this situation, the overburden thickness increases rapidly as the trench advances into the hillside, eventually resulting in an uneconomic **stripping ratio** of overburden to valuable rock. Mining activity then moves laterally along the outcrop, approximately parallel to the ground contours in a subhorizontal deposit. Small-scale contour strip mining, particularly of coal and Carboniferous ironstones, was common in Britain until the nineteenth century. Its effects are minor compared with those in areas such as the Appalachian coalfields of the USA.

Shallow open-cut extraction is typical of near-surface, poorly consolidated deposits with a moderate to low waste ratio (Fig. 14.6d). Examples are sands and

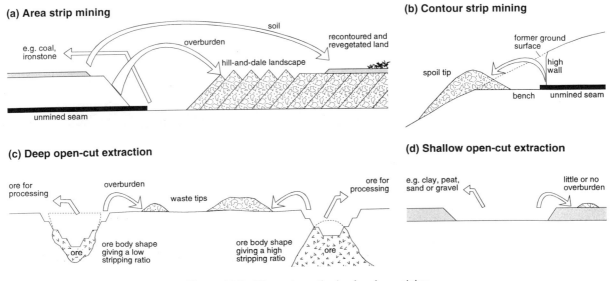

Figure 14.6 The main methods of surface mining.

gravels, peat, or shallow brick clays. In contrast to strip-mining, the overburden or processing wastes are insufficient to backfill the excavations. If open-cut pits reach below the water table, they will flood when extraction and pumping cease. The conversion of land into lakes was seen as environmental degradation, until the recreational and wildlife potential of some of these lakes was realized. Nevertheless, large areas of unreclaimed sand and gravel workings remain in the British Isles, for instance along the valleys of the Thames and its tributaries. The brick pits of the Oxford Clay (Ch. 8) provide further examples of the impact of open-cut extraction. In the Bedfordshire brickfield (Fig. 14.7) about 15 metres of brick clay underlie a maximum of 10 m of overburden, so that abandoned pits are either flooded or part filled to above the water table with ridges of waste.

Offshore dredging for sand and gravel resources is a special form of open-cut extraction, altering the shape of the seafloor and the pattern of marine currents in ways that can aggravate coastal erosion.

Open-cut extraction above the water table does not result in flooded workings, and some sites can be re-instated with soil and vegetation. An exception is where the soil cover itself is extracted. The main example in the British Isles is the stripping of peat for fuel and horticulture. Together with drainage and development, this extraction has degraded over 90% of the region's original peatland habitats. The impacts are particularly serious in Ireland, where peat supplies nearly a fifth of primary energy demand.

Deep open-cut extraction is used for irregular or steeply dipping deposits, typically of industrial or metallic minerals. The amount of waste depends on the grade and shape of the deposit (Fig. 14.6c). Overburden must be removed from most open-cuts simply to maintain the stability of their slopes. In contrast to strip mines, this overburden must be dumped beyond the planned area of the excavation, increasing the environmental impact of deep open-cut working. The most extensive examples in Britain are the china clay pits of southwest England (Fig. 14.8), which produce around ten times as much waste quartz and mica as usable kaolinite. Dumping this waste in disused pits would prevent their future exploitation. Instead, waste tips occupy much of the land between the pits, causing widespread degradation of the landscape.

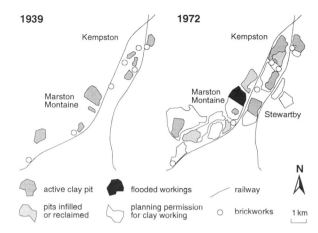

Figure 14.7 Changes in the area of brick clay workings in part of Bedfordshire between 1939 and 1972.

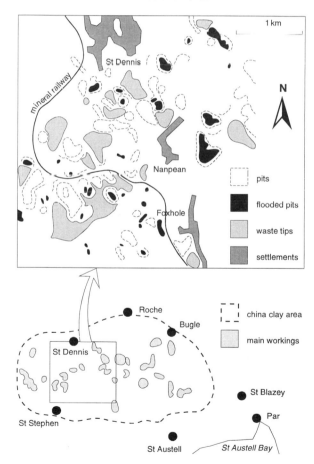

Figure 14.8 Land use in part of the main area of china clay working in Cornwall.

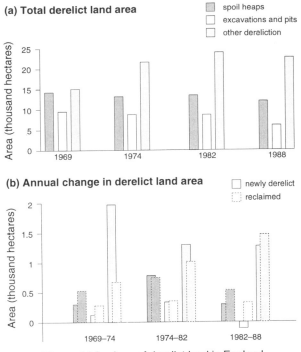

(a) Total derelict land area

spoil heaps
excavations and pits
other dereliction

(b) Annual change in derelict land area

newly derelict
reclaimed

Figure 14.9 Area of derelict land in England.

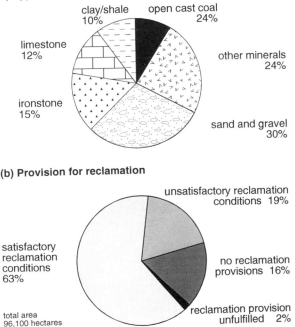

(a) Type of mineral working

clay/shale 10%
open cast coal 24%
limestone 12%
other minerals 24%
ironstone 15%
sand and gravel 30%

(b) Provision for reclamation

unsatisfactory reclamation conditions 19%
satisfactory reclamation conditions 63%
no reclamation provisions 16%
reclamation provision unfulfilled 2%

total area 96,100 hectares

Figure 14.10 Surface mineral working in England in 1988.

14.5 Impacts in the British Isles

The environmental impact of extracting geological resources can be gauged from surveys in England of **derelict land**, so damaged by human activity as to need remedial treatment before further use. Mineral extraction is responsible for more derelict land than any other single activity (Fig. 14.9a). In 1988, spoil heaps made up 30% of the derelict areas, and excavations about 15%. The "other dereliction" arises mostly from general industrial, railway and military use, although mining subsidence accounts for a further 2.5% of the total (Fig. 6.8).

Extraction-related dereliction has decreased steadily from an estimated 25 000 ha (64% of the total) in 1969 to 19 000 ha (47.5%) in 1988. Land is therefore being reclaimed faster than it is produced by the closure of pits, mines and quarries (Fig. 14.9b). However, the net annual reclamation is only a few per cent of the total derelict workings, and at this rate it will be over 25 years before all this land is reusable. New extraction permissions are subject to more stringent restoration conditions than were old licences. However, inadequate reclamation conditions still apply to over a third of the 96 000 ha of permitted surface workings in England, mostly for construction materials (Fig. 14.10). These sites will add to the stock of derelict land when present working finishes.

Underground mining permits affect at least eight times the area of surface licences, but do require compensation for subsidence damage. Not covered is the water pollution from **acid mine drainage**. This is the acidic, usually metal-rich outflow of the groundwater which permeates abandoned workings once pumping has ceased. The rapid closure of coal mines in Britain is likely to exacerbate such pollution problems.

Tightening planning regulations still lag behind rising standards of public environmental awareness. Now "restoration" of land to agricultural or recreational use is seen as inadequate if the pre-existing habitat or scenery resulted from centuries of natural processes. The example of peatlands has already been mentioned (Ch. 14.4). Another case is the extraction of crushed rock aggregate, requiring extension of existing quarries in scenically attractive areas or opening of large **coastal superquarries** in remote areas of Ireland and Scotland (Ch. 8.4).

14.6 Global perspectives

Many of the impacts of resource extraction in the British Isles are replicated worldwide. However, added problems occur in some countries because of the nature or scale of the geological resource. Over 30 billion tonnes of rock are now moved each year in extracting non-fuel minerals and rocks alone, twice the annual sediment load of the world's rivers.

Uranium ore is one of the resources that presents special problems. Released radon gas is a radiation hazard in underground mines, but quartz dust is a greater threat in open-cut workings. The highest radiation risk is in processing areas, where ore is roasted to recover uranium oxide. The public is at risk from ingesting dust or water from radioactive waste tips at abandoned mines. The worst sites identified so far are in southeast Germany, the world's third largest uranium-producing region.

Gold extraction involves processing large volumes of rock to extract its few parts per million of metal. **Hydraulic mining**, washing out rock with high-pressure water jets, has been particularly damaging in countries such as Indonesia, the Phillipines, and those of the Amazon basin. Rivers and lakes become badly silted, and also polluted by the mercury used in separating the gold. A further pollution risk in North America is the cyanide used to leach gold from heaps of crushed low-grade ore or tailings. The recent issue of hundreds of licences for gold exploration in Scotland and particularly in Ireland have now made gold extraction an environmental issue in the British Isles.

Open-cut copper mines have the world's largest excavations, and therefore pose extreme problems of waste disposal. For instance, the Panguna mine on the island of Bougainville (Papua New Guinea) dumped 130 000 tonnes of metal-contaminated tailings each day into the adjacent river, covering 1000 ha with waste and destroying aquatic life for 30 km downstream. This environmental destruction triggered a civil war that closed the mine in 1989.

The indirect impacts of mineral extraction are often more serious even than those of excavations or waste tips. Smelting of metallic ores produces about 8% of the global emissions of sulphur dioxide – the primary cause of acid rain (Ch. 17) – together with arsenic, lead, cadmium and other heavy metals. Notorious environmental "dead zones" have been produced around smelters such as Sudbury and Trail (Canada), Copper Hill (Tennessee, USA) and Severonikel (Russia). In the Amazon basin of Brazil, the supply of fuelwood to the Carajas iron ore mine and smelter threatens to destroy about 50 000 ha of forest each year.

The worldwide market in geological resources, particularly in metallic ores and fossil fuels, turns these local impacts into global concerns. A serious accusation is that the consumer nations of the industrialized world avoid paying the real costs of the environmental impacts suffered by less wealthy producer nations. Such issues of **inter-regional equity** will be confronted again in Chapter 18.

Further reading

Brown, J. R. 1992. *State of the world 1992*. London: Earthscan. [Chapter 7, by J. E. Young, is a review of impacts of mining worldwide.]

Department of the Environment 1991. *Survey of derelict land in England 1988*. London: HMSO. [The latest of the periodic government surveys of the extent and type of derelict land.]

Goudie, A. 1990. *Human impact on the natural environment*, 3rd edn. Oxford: Basil Blackwell. [Covers the impacts of extraction in the wider context of human effects on landscape.]

Irish Peatland Conservation Council 1992. *Policy statement and action plan*. Dublin: Irish Peatland Conservation Council. [Guide to the environmental impacts of peat extraction.]

Lumsden, G. I. (ed.) 1992. *Geology and the environment in Western Europe*. Oxford: Oxford University Press. [Chapter 5 contains European examples of the impacts of resource extraction.]

O'Dowd, L. 1990. *Gold mining and the Irish environment*. Bantry, Co. Cork: Earthwatch. [A survey of the potential impacts of mining gold and extracting it by leaching.]

Smith, P. J. (ed.) 1975. *Politics of physical resources*. Harmondsworth: Penguin. [Contains well-documented case-studies of extraction impacts in the UK.]

Wallwork, K. L. 1974. *Derelict land*. Newton Abbott: David & Charles. [Comprehensive survey of the causes and remedies of derelict land, with case studies of mines and quarries.]

Waltham, A. C. 1989. *Ground subsidence*. Glasgow: Blackie. [An excellent introduction to both natural and mining-induced ground subsidence.]

Chapter Fifteen

ENGINEERING GEOLOGY

15.1 Physical principles

Many engineering projects are attempts to exploit, modify or augment the natural environment for the benefit of human society. The subsurface geological environment – particularly the state of stress and water flow – is perturbed wherever the ground is excavated or structures built upon it. Good engineering practice involves predicting and minimizing these subsurface impacts, both the transient effects during construction work and any long-term consequences. Negligence can threaten the safety of both the engineered structure and its surroundings.

The geological parameters relevant to **engineering geology** (Fig. 15.1) fall into two main categories – those concerned with the **strength** of rocks or soils and those describing the behaviour of **subsurface fluids**. Some of these parameters can be determined by laboratory testing on a microscopic or hand-specimen scale, but many can be measured only on the site of interest.

Rock strength is determined primarily by the mineralogical nature of the grains, their shape, proportions and mutual arrangement. All rocks have a **co-efficient of friction** – a resistance to sliding along a plane through the rock – due to the surface roughness and interlocking of grains. Some rocks also have a **cohesion** due to a secondary mineral cement or to electrostatic forces between small clay grains. The strength of a rock is affected by its **porosity** – the proportion of voids between the grains – which also determines the maximum fluid content. The interconnectedness of pores determines the rock **permeability** – the rate at which pore fluids can flow through.

Bulk rock strength can be determined directly by laboratory testing. Both **shear strength**, dominantly under compression, and **tensile strength** may be relevant. Rock strength must be greater than the sum of the *in situ* stress measured in the working area, the additional stresses imposed by new structures, and any dynamic stresses predicted from earthquakes. Rock strength is lowered by a high **fluid pressure** within the pores, itself sometimes caused by the compaction of weak impermeable rocks by the weight of a newly built structure. Most important, the strength of large masses of rock is controlled as much by any weak bedding planes and fractures that it contains as by the inherent strength of the intact rock.

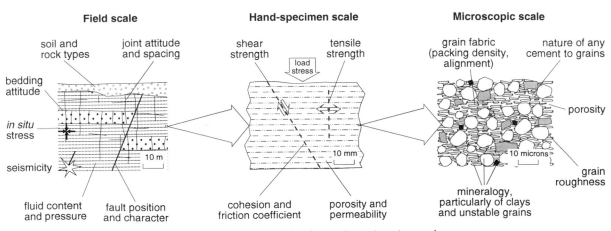

Figure 15.1   Some geological factors in engineering geology.

15.2 Foundations

The **load** of a built structure is transferred to the ground through engineered **foundations**. A range of foundation problems can occur, depending on the nature of the underlying geology (Fig. 15.2).

All structures undergo some vertical **settlement** by the elastic deformation of the loaded rocks. However, some materials, particularly clays, are prone to large permanent **compaction** as pore fluids are expelled and grains rearranged. Also damaging to structures is **ground swelling**, caused by clay or anhydrite beds that expand on wetting, sometimes as the water table is perturbed by building development itself.

Soils or weakly cemented rocks tend to deform by **rotational failure** if their brittle strength is exceeded by the imposed load. Deformation is concentrated on a near-circular failure surface, and some rotation of the affected structure is common. In stronger and well bedded rocks, **sliding failure** is favoured. The failure surface tends to follow a weak bedding plane or an interface between strong basement and weaker cover.

One solution to foundation problems is to spread the load over a wider area with concrete **pads**, **strips** or **rafts** below columns, walls or the whole structure. The other solution is to sink **piles** through weak shallow materials down to strong rocks at depth. Small structures can be built on most substrates by one of these methods. However, the design and siting of large

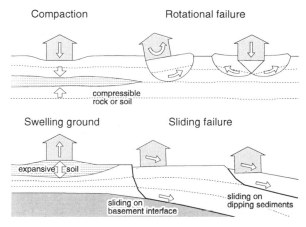

Figure 15.2 Types of foundation problem.

structures may be strongly controlled by the available geology. The road bridge across the Severn Estuary north of Bristol is an example (Fig. 15.3). A suspension bridge was required by a large tidal range and a deep channel at the narrowest crossing. The southeast pier and suspension anchorage were sited on strong Carboniferous limestones exposed near low-water mark. The northwest anchorage could be founded on limestone below 18 m of excavated weak overburden. The limestone beneath the northwest pier is impractically deep, but fortunately the consolidated Triassic mudstones proved sufficiently strong at about 10 m below their outcrop surface.

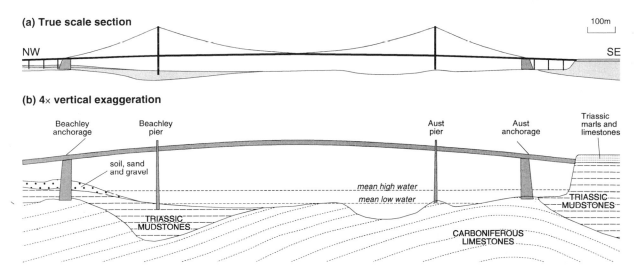

Figure 15.3 Geological foundations to the road bridge across the Severn estuary between Gwent and Avon.

15.3 Excavations and embankments

Many engineering projects alter the natural ground contours by digging out rock or soil and dumping it elsewhere. In particular, the fairly even gradients required for roads, railways and canals necessitate **cuttings** through high ground and **embankments** over low ground. These new landforms may be unstable without proper engineering of their bounding slopes. Stabilization is more important still in underground excavations such as tunnels and storage caverns.

The slopes of a surface excavation may fail by **slumping** of weak materials or by **sliding** or **toppling** of stronger rocks (Fig. 15.4a). Planar weaknesses such as bedding or fractures determine the failure risk in hard rocks. Sliding is a hazard where weak planes dip gently into the excavation on an **up-dip face**. A **down-dip face** is more stable, but may suffer toppling failure. Tunnels are prone to unwanted enlargement or **overbreak** of their profile, by **falls** from the roof and **bursts** from the walls. The risk of bursts in strong rock and **ground squeezing** of weak rock is increased if the subsurface compressive stress is high.

The engineering remedies for unstable excavations involve **ground treatment** or **support** (Fig. 15.4b). **Grouting** fills rock fractures and pores with cement, bitumen or chemical precipitates, in order to decrease permeability and increase strength. Unconsolidated sediments can be temporarily strengthened during excavation by **dewatering** or **freezing** them. Some slopes can be permanently strengthened by **rock bolts**

(a) Engineering problems in excavations

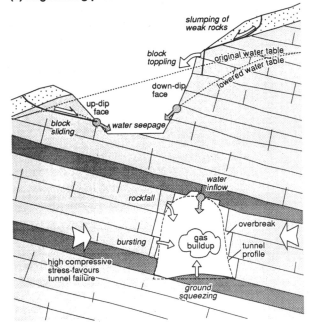

(b) Stabilization of excavations

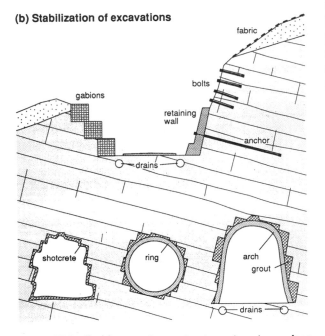

Figure 15.4 Problems and remedies in engineering surface and underground excavations.

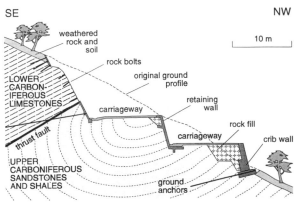

Figure 15.5 Engineering of rock slopes on the M5 motorway across the Cleveden Hills, Avon. From Eyre (1973) with the permission of the Geological Society Publishing House.

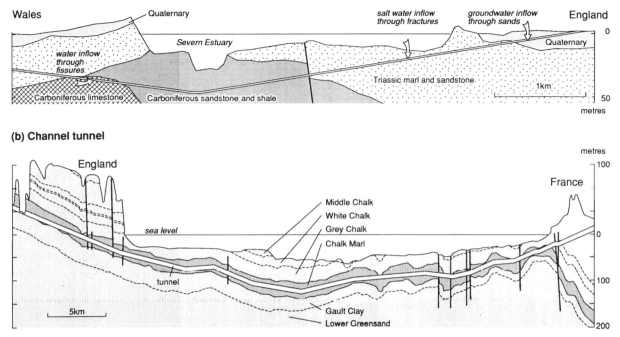

Figure 15.6 Geological cross sections of the Severn and Channel tunnels.

grouted into the face or mesh **fabrics** draped over the slope. Weak materials may need a **retaining wall** of concrete, rock or **gabions** – mesh cages filled with stones. Most tunnels require supporting with **arches** or **rings** of concrete or steel, although a sprayed concrete lining can be adequate in strong rocks.

All excavations disturb the flow of groundwater (Fig. 15.4a). Tunnels in permeable rocks below the water table need to be pumped during excavation and provided with a permanent waterproof lining. Surface excavations into the water table promote inflow which may progressively destabilize the bounding slopes. The overall lowering of the water table may cause nearby wells and springs to dry up.

The expansion of Britain's motorway system has caused serious environmental impacts, often minimized only by difficult engineering solutions. For example, the M5 motorway had to cross the Clevedon Hills, southwest of Bristol, along a slope containing a weak zone of thrust faulting (Fig. 15.5). Building the two carriageways at different levels reduced the volume of the excavation and avoided a high faulted face. The fault zone was supported by a retaining wall an-

chored into unfaulted rock. The overlying limestone face was stabilized by rock bolts.

Geological factors in planning underground excavations are well illustrated by comparing two long railway tunnels, built over a century apart. The Severn tunnel (Fig. 15.6a), completed in 1886, was started without any subsurface exploration, although five shafts and a borehole were sunk as tunnelling proceeded. Severe problems arose from groundwater inflow, through Quaternary sands, joints in the Triassic, and most seriously through solution cavities in the Carboniferous limestones. Despite later grouting work, this inflow continues to be a maintenance problem. By contrast, the construction of the Channel Tunnel during 1986–94 was preceded by extensive borehole and seismic surveys. A generally synclinal structure allows about 85% of the tunnel to run within the Chalk Marl. This is an ideal tunnelling material, stronger than the underlying Gault but without the troublesome fracture permeability of the purer Chalk above. In only a few folded or faulted zones can the 8 m diameter running tunnels not be contained within the 25 m thick marl.

15.4 Water engineering

The primary rôle of some engineering schemes is to intervene in the natural hydrological cycle. The aim may be to supply water or hydropower, to provide flood protection or reclaim wetlands. The main strategies are **channelization** of rivers and their **damming** to form **reservoirs** (Fig. 15.7). Other human activities have secondary impacts on hydrology – the management of coastlines (Ch. 15.5), the development of urban areas (Ch. 15.7), and the modification of land by agriculture and forestry (Ch. 15.6).

Channelization of rivers involves a spectrum of increasingly invasive engineering works. Least perturbing is local **bank protection** of an existing channel to save bank-side structures from erosion damage. Protective structures include lines of **piles**, and **retaining walls** of concrete, armourstone or gabions. These defences may coalesce so that rivers become **conduited** in an engineered channel, often with raised banks or **levees** to prevent overspill of floodwater. A more severe development is **diversion** of a river into an artificial channel. Straightened channels free land

for development in urban areas. They have been used to drain wetlands such as the Somerset Levels and the East Anglian Fens (Ch. 15.6) for agricultural use, the former floodplains being kept dry by pumping water into the new channels. **Canals** are channels engineered primarily for navigation. The need to maintain a nearly level course means that, unlike natural channels, canals tend to run along rather than across the ground contours.

Although channelization has beneficial aims, it can have harmful and often unplanned impacts. The extra capacity of a smoothed river channel causes increased bank erosion further downstream. Subsidence of former flood plains results from a lowered water table and the starvation of flood-borne sediment. Bypassed and drained wetlands may have greater value as natural habitats than as agricultural or urban land.

Rivers in the British Isles have been extensively channelized (Fig. 15.8). In England and Wales the proportion varies from about 12% in northwest England to over 40% in the London area. However, the resulting impacts are moderated by the relatively small size and sediment load of the rivers involved.

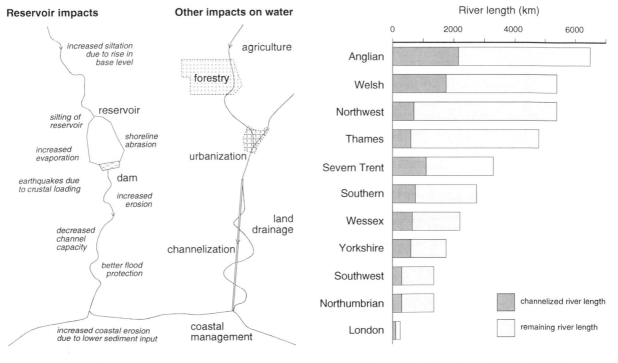

Figure 15.7 Some impacts of water management.

Figure 15.8 River channelization in Britain.

Dams offer a major engineering challenge, particularly because the potential consequences of their structural failure are catastrophic. The design of a dam depends on the strength of the bedrock and the availability of construction materials (Fig 15.9). An **embankment dam** has an impermeable core of clay or concrete supported by wide shoulders of soil or rock fill. These dams spread their weight over a large basal area and need no strong foundation, but do require large volumes of fill. **Concrete dams** need strong bedrock that will not settle much under load. **Arch dams** use least concrete, and gain strength from their curved shape both in section (Fig. 15.9) and, most important, in plan. **Gravity dams** depend on the weight of their massive triangular section. **Buttress dams** have a series of watertight slabs supported on spaced buttresses.

Any dam is liable to leak through the surrounding bedrock. To reduce this an impermeable barrier or **cut-off** is formed in the foundations of the dam and in its abutments onto the valley walls (Fig. 15.9), usually by injecting a curtain of grout. Even then some dams have leaked severely, usually due to inadequate knowledge of their bedrock. The Dol-y-gaer dam in South Wales was built at a narrow point where faulted slivers of Carboniferous limestone cross the Taf Fechan valley (Fig. 15.10). Severe leakage occurred through fractures. Remedial measures failed, and a new dam had to be built down stream, founded on less permeable Devonian mudstones.

Dams and their reservoirs have even greater environmental impacts than does channelization (Fig. 15.7). Valleys up stream of the reservoir may aggrade by deposition as their streams adjust to the new base level. In strongly eroding areas the reservoir itself will silt up. Decrease in streamflow below the reservoir alters the stream profile, and may affect downstream irrigation and water supply. The reservoir encroaches on formerly useful land, hosts increased evaporation from the drainage basin, changes the local microclimate, and may generate earthquakes due to crustal adjustment to the load of the water. Such impacts have to be weighed against the benefits of water supply, flood control or power generation. In Britain, the balance is changing, such that the cost of conserving water is arguably less than the environmental impact of impounding more supplies.

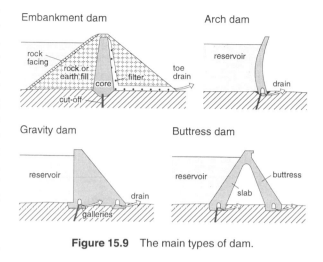

Figure 15.9   The main types of dam.

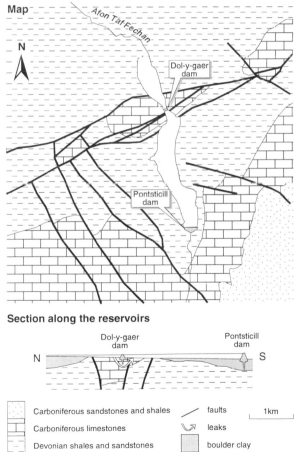

Figure 15.10   The geological setting of the Taf Fechan dams, South Wales. © Crown copyright.

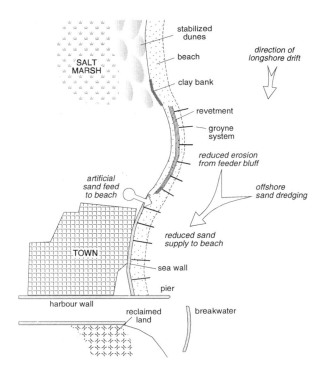

Figure 15.11 Coastal management strategies.

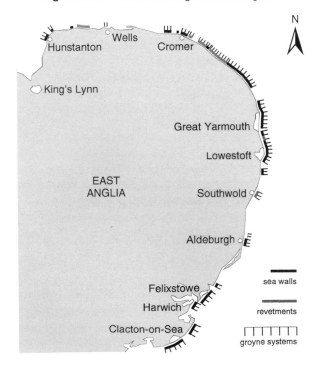

Figure 15.12 Coastal defences in East Anglia.

15.5 Coastal management

Much human activity in the British Isles is concentrated near its coastline, many segments of which are tending to advance or retreat (Ch. 4.5). Engineering structures have been variously used to minimize coastal erosion, to prevent flooding or to encourage coastal advance by reclaiming land (Fig. 15.11). These aims are not always mutually compatible.

The most direct coastal protection is a **sea wall** built close to high water. However, this reflects rather than absorbs wave energy, often enhancing erosion to seaward and requiring artificial **sand feeding** to maintain the beach level. A **clay bank** is a softer, cheaper alternative, although more prone to overtopping by waves. Also absorbent to wave energy is a slatted **revetment** built below high water, designed to trap sediment and cause less seaward erosion. However, beach sand may still be lost without **groynes** constructed at right angles to the coastline. These wooden or concrete fences slow the longshore drift of sand. **Piers**, **breakwaters** and **harbour walls** also interfere with natural sediment flows, although they are not built primarily for that purpose.

Tidal flats and salt marshes can be reclaimed for agriculture by surrounding them with **flood barriers** such as banks or walls and draining them either naturally or by pumping. Land levels can be raised for building by dumping rock fill or solid waste.

The environmental impact of coastal defences has increased steadily over past decades. About three-quarters of the East Anglian coastline is artificially protected (Fig. 15.12). Many defences were built or strengthened after the 1953 storm surge (Ch. 4.6) and now need repair or replacement at costs of between £1 m and £5 m per kilometre. Along a rural coastline this cost is commonly more than the value of the agricultural land being defended. An alternative though controversial strategy of **managed coastal retreat** may be necessary. Houses on vulnerable land would be relocated and sea defences breached. Cliffed coasts would be free to erode and on low-lying coasts the beach/dune/salt marsh system would be allowed to re-establish as the natural buffer to wave energy. Maintenance of this system depends on an adequate supply of sediment, severely reduced on an engineered coastline by preventing **feeder bluffs** from eroding.

15.6 Agriculture

Some farming and forestry practices have the effect, often unintentional, of reshaping the land surface. In this respect, agriculture is a form of land engineering.

The most widespread environmental impact from agriculture is the increased risk of soil erosion when land is deprived of a permanent cover of vegetation. About a third of the arable area of England and Wales is thought to be at risk (Fig. 15.13). Wind erosion particularly affects drained peat soils or areas of Quaternary sands. Water erosion is a threat where sandy soils are used for arable farming and where blanket peats in upland areas are disturbed.

Practices that increase the risk of soil erosion are the ploughing of former pasture or moorland for either arable crops or forestry, the removal of field boundaries, and the use of autumn-sown crops which expose bare ground to winter and spring rains. Annual soil losses of up to 18 tonnes per hectare have been measured on sloping bare soils subject to water erosion, nearly ten times the rate from grassland. Enhanced erosion from natural woodland is negligible.

Ploughing parallel to contours reduces water erosion and the replanting of hedges or tree shelter-belts also protects land from wind.

More localized but more obvious environmental impacts arise from the draining of estuarine marshes and fens for agricultural use. The East Anglian Fens were progressively drained from the mid-17th century onwards by cutting wide straight channels to replace the naturally flood-prone meandering rivers (Fig. 15.14). However, the floodplain sediments shrank as they dried out, particularly the peat soils of the southern Fenland (Fig. 15.14 inset). The main drainage channels had to be kept raised by embankments, and drainage water had to be pumped up into them from the surrounding fens. The surface of the peat fen is now about 5 m below its pre-drainage level, with large areas at or below sea level.

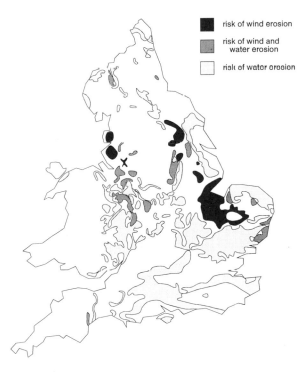

Figure 15.13 The soil erosion risk in England and Wales.

risk of wind erosion
risk of wind and water erosion
risk of water erosion

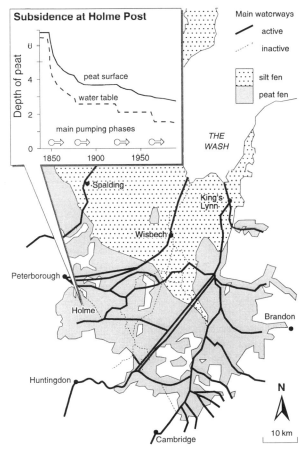

Figure 15.14 Subsidence and drainage in the Fens.

15.7 Urbanization

About 87% of people in the British Isles live in urban areas, a quarter of these in cities of more than a million inhabitants. These towns and cities have been growing throughout the 19th and 20th centuries, resulting in progressive **urbanization** of the landscape. Some of the engineered components of this process have already been mentioned – buildings, roads, tunnels, channelized rivers. However, urbanization as a whole has its own suite of environmental impacts, many involving perturbation of surface water and of groundwater (Fig. 15.15).

An early effect of urban growth is that rain, previously absorbed by vegetated land, runs off the impermeable surfaces of buildings and roads. The reduced infiltration, together with pumping of groundwater from wells, lowers the underlying water table. Water that does reach the subsurface is increasingly polluted by domestic and industrial effluents. Reshaping of the land for foundations combines with the increased runoff to promote soil erosion.

Urban growth involves channelization of rivers and progressive development of their floodplains. The greater runoff into these constrained channels increases the risk of overbank flooding after heavy rains while reducing the **base-flow**, the fair weather throughput of water. The flood risk is increased fur-

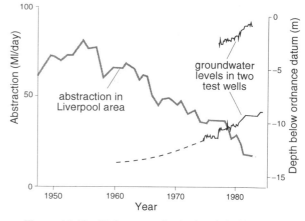

Figure 15.16 Rising groundwater levels in Liverpool.

ther by natural shrinkage of the dry floodplains and their subsidence under the load of large buildings.

In later stages of urban growth, the local groundwater sources may be abandoned due to pollution and a lowered water table. Water is piped into the city from distant sources. This substitution of imported water allows the water table to rise towards its pre-urban level. Rising water tables have been recorded in Liverpool and other British cities as local abstractions have fallen (Fig. 15.16). Such rises threaten the stability and drainage of tunnels and foundations built during the time of lowered water tables.

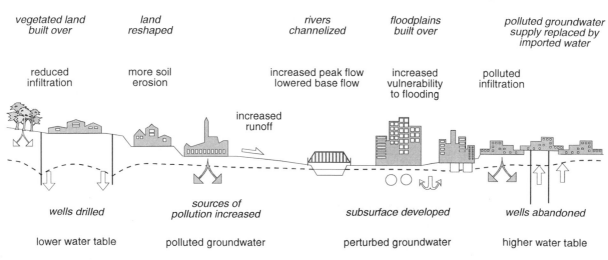

Figure 15.15 Some effects of progressive urbanization.

15.8 Global perspectives

The engineered environment in the British Isles is typical only of the developed world. Other regions have a different balance of engineering impacts.

Urbanization is one index of these contrasts. (Fig. 15.17). In Europe, North America, Oceania and the former USSR, urban areas are growing annually at 2% or less, and at least two-thirds of the population now live in towns and cities. The urban growth rate in South America, Asia and Africa is about double this. In South America 75% of the population now live in urban areas. The long-established mega-cities of New York, Tokyo, London and Paris have been matched or exceeded in size by places such as São Paulo, Mexico City, Shanghai, Buenos Aires and Calcutta. The size and rapid growth of these cities present formidable problems for planners and engineers. Indeed, it is estimated that between 70% and 95% of new housing in Third World cities is unplanned and lacks a properly constructed infrastructure for transport, water and sewage.

The global average length of paved road per 1000 km² of land area is about 90 km (Fig. 15.17). By contrast, the density in North and Central America is about double this, in Europe about eight times and in the UK over 16 times this average density. New roads in less developed countries often present engineering problems, sometimes in hostile terrain. However, these problems are compounded less often by the conflicts of land-use which invariably arise in extending an already invasive road network such as that in Britain.

One type of engineering that impacts heavily on the developing world is the construction of dams and reservoirs, mainly for hydroelectric power schemes. About two-thirds of the world's large dams are in Asia, over three-quarters of these in China alone. Some projects are controversial because they destroy natural habitats, displace populations, disrupt hydrology, and pose an extra hazard in seismically active areas. An even more potent environmental threat is the "land engineering", often unintended, caused by deforestation and other agricultural practices. About 17% of the world's vegetated land is reckoned to be degraded, much of it through erosion caused by human activity (Ch. 6.5).

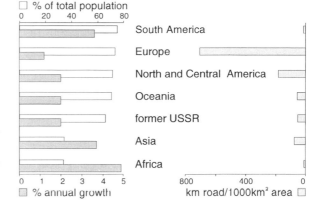

Figure 15.17 Measures of engineering development in different world regions.

Further reading

Anderson, J. G. C. & C. F. Trigg 1976. *Case-histories in engineering geology.* London: Paul Elek. [A useful collection of case-studies, dominantly from the British Isles.]

Blyth, F. G. H. & M. H. de Freitas 1984. *A geology for engineers,* 7th edn. London: Edward Arnold. [A long-established introductory text, still one of the most informative available.]

Coates, D. R. 1981. *Environmental geology.* New York: John Wiley. [Part four deals with the human modification of nature, and includes an excellent overview of engineering geology and urbanization.]

Darby, H. C. 1969. *The draining of the fens,* reprint. Cambridge: Cambridge University Press. [The classic account of the history of drainage engineering in the East Anglian Fens.]

Legget, R. F. & A. W. Hatheway 1988. *Geology and engineering.* New York: McGraw-Hill. [A North American view of engineering geology, but with a worldwide collection of case studies.]

Lumsden, G. I. (ed.) 1992. *Geology and the environment in Western Europe.* Oxford: Oxford University Press. [Section 3.6 is a useful assessment of underground space as a resource and section 5.2 a discussion of the impacts of development.]

McLean, A. C. & C. D. Gribble 1985. *Geology for civil engineers.* London: Allen & Unwin. [A reliable textbook with some case-studies from the British Isles.]

Chapter Sixteen

LAND AND WATER POLLUTION

16.1 Wastes and pollution

Pollution is the disturbance of the Earth's life-support systems, most critically those that recondition land, air and water resources. The most publicized of pollution incidents affect the atmosphere or surface waters. However, the subsurface environment is also prone to pollution, particularly of its groundwater. Contamination here is slower to spread but slower to remedy than that influenced by the more rapid processes on the Earth's surface.

Pollution is due to **wastes** from society being released at rates and concentrations too high for natural systems to assimilate. Some types of waste are recognized by society as potential environmental threats, unless they are properly collected and disposed of. These **managed wastes** have a known disruptive effect on natural systems, either through their chemical or biological toxicity or their physical bulk (Fig. 1.16b). They contrast with **unmanaged wastes** (Fig. 1.16a). Some of these wastes are accidental leakages or spillages of known toxic materials. Most are diffuse releases of substances that either do no proven harm, or do damage only to a limit which society is prepared to tolerate. The line between managed and unmanaged wastes is continually under review as more substances are implicated in environmental damage. The philosophical basis of pollution control is also in question. Should a particular type of waste be considered safe until proved otherwise, or should all wastes be suspect until shown to be harmless – the so-called **precautionary principle** (Ch. 18)?

The blurred distinction between safe and hazardous wastes is exemplified in modern agriculture. About 250 million tonnes of excrement is produced annually by farm animals in the UK. About 20% of this is from intensively reared livestock, and is accounted as managed waste because it must be collected and either treated or spread as a fertilizer (Fig. 16.1b). However, most excrement is dispersed by free-grazing animals themselves. This diffuse, unmanaged waste (Fig. 16.1a) is less likely to cause intense pollution. Indeed, it was a vital and beneficial input within traditional farming rotation systems. More serious pollution comes from the unmanaged residues of fertilizers, pesticides and herbicides, designed to replace and augment traditional agricultural methods.

Some other sources of unmanaged waste that can threaten the subsurface environment are untreated run-off from roads and paved areas and acid deposition from dissolved atmospheric gases (Fig. 16.1a).

Although managed wastes make up only part of the total pollution threat, their sources and disposal routes are the better monitored (Fig. 16.1b,c). Some disposal strategies aim to **dilute and disperse** pollutants in a large-volume medium. Examples are the dumping at sea of chemical wastes and the spreading on farmland of animal wastes and sewage sludge. The capacity of the sea to neutralize pollutants has often been over-estimated, so that sea dumping is increasingly discouraged. Landfill sites, the most common waste disposal method in the UK and Ireland, also used to operate on this dilute-and-disperse principle, with liquid effluents being allowed to leak into groundwater. New landfills now aim to **concentrate and contain** the toxic fluids that are leached from them. These leachates, together with other hazardous industrial chemicals, must be chemically treated, or incinerated. Incineration produces its own wastes, many as gases which in turn can overload the dilution capacity of the atmosphere (Ch. 17).

This chapter is specifically concerned with the pollution of land and groundwater by wastes. **Point-source pollution** (Ch. 16.2) arises from leaks and spills at discrete sites. **Diffuse pollution** (Ch. 16.3) arises from more widespread influx of wastes, for example from agriculture or acid rain. Radioactive waste (Ch. 16.4) is an example of a hazardous material needing a particularly stringent containment strategy.

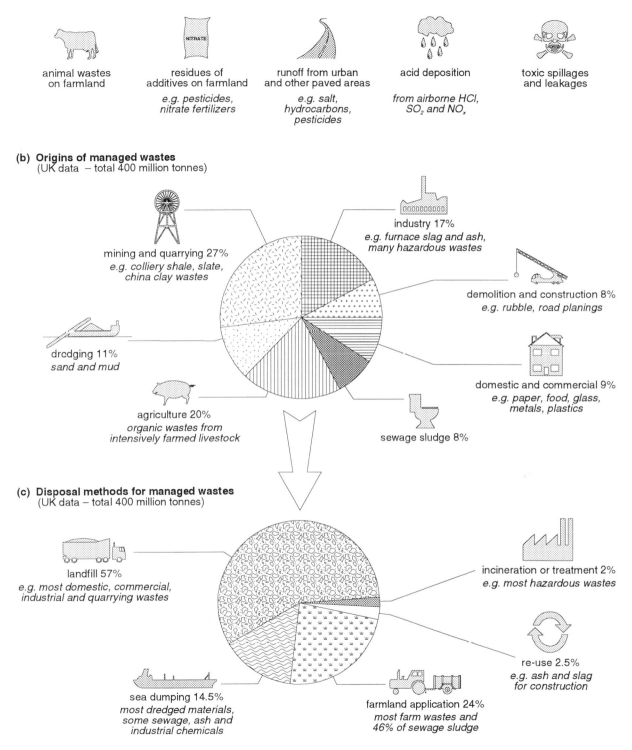

(a) Unmanaged wastes

animal wastes on farmland

residues of additives on farmland
e.g. pesticides, nitrate fertilizers

runoff from urban and other paved areas
e.g. salt, hydrocarbons, pesticides

acid deposition
from airborne HCl, SO₂ and NOₓ

toxic spillages and leakages

(b) Origins of managed wastes
(UK data – total 400 million tonnes)

mining and quarrying 27%
e.g. colliery shale, slate, china clay wastes

industry 17%
e.g. furnace slag and ash, many hazardous wastes

demolition and construction 8%
e.g. rubble, road planings

dredging 11%
sand and mud

domestic and commercial 9%
e.g. paper, food, glass, metals, plastics

agriculture 20%
organic wastes from intensively farmed livestock

sewage sludge 8%

(c) Disposal methods for managed wastes
(UK data – total 400 million tonnes)

landfill 57%
e.g. most domestic, commercial, industrial and quarrying wastes

incineration or treatment 2%
e.g. most hazardous wastes

re-use 2.5%
e.g. ash and slag for construction

sea dumping 14.5%
most dredged materials, some sewage, ash and industrial chemicals

farmland application 24%
most farm wastes and 46% of sewage sludge

Figure 16.1 The main sources of, and disposal strategies for, wastes that can pollute land and water.

127

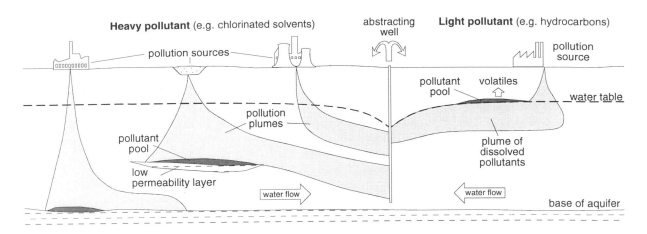

Figure 16.2 Subsurface behaviour of pollution from point sources.

Map

Section AB

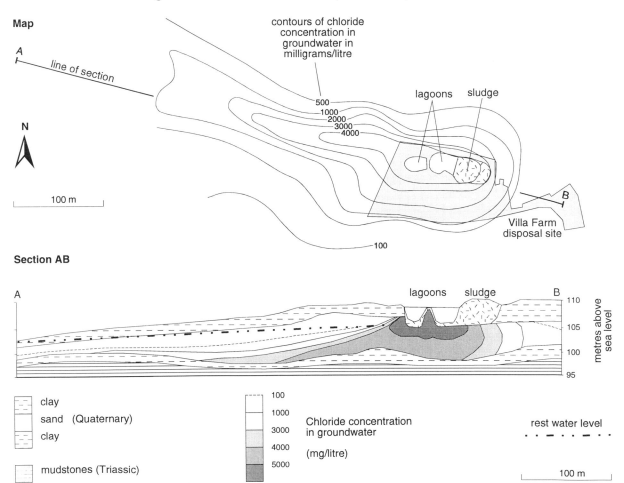

Figure 16.3 Example of a pollution plume from liquid waste lagoons near Coventry. Reproduced from QJEG (Williams et al. 1984) with the permission for the Geological Society Publishing House.

16.2 Point-source pollution

Some pollution of groundwater can be traced to one or more discrete sources of contaminant. Examples of active sources are industrial sites such as chemical or engineering works, unconfined landfills, and soakaways from paved areas. However, the generally slow response of groundwater systems means that present pollution often results from the past use of a site. Abandoned mines can leak acidic water, and mine tips produce leachates rich in chloride, sulphate and metals. Old landfills continue to leak for many decades. The sites of former gasworks are contaminated with tars and other processing residues.

Pollutants from a point on the Earth's surface spread downwards and laterally to form a discrete **plume** of contamination at depth (Fig. 16.2). The form of this plume depends on whether the pollutant is **miscible** with water and on its density relative to water. Low-density fluids such as light hydrocarbons form a floating **pool** at the water table. Volatile components evaporate to the surface from this pool. Soluble components generate a plume at the top of the saturated zone, which migrates laterally in response to regional water flow.

Pollutants with a density greater than water continue to sink through the water table, with their more soluble components progressively dissolving. The plume spreads laterally with the regional groundwater flow or when it reaches less permeable rocks within or below the aquifer. Undissolved pollutants may form a pool above these impermeable layers.

A well studied pollution plume resulted from contamination of a Quaternary sand aquifer by a disposal site for industrial liquids and sludges (Fig. 16.3). Wastes were dumped into a lagoon from which floating oil was skimmed for reprocessing and settled sludge was removed to a landfill site. The remaining aqueous fluid was periodically pumped into tankers for sea disposal. However, an estimated $7000 \, m^3/year$ of this fraction infiltrated the aquifer between 1967 and 1982. The resulting plume was defined by sampling the chloride concentration in test boreholes around the site. The plume is aligned with a westward hydraulic gradient and localized by a west–northwest trending channel at the base of the sand.

Contaminated groundwater is difficult to treat, so that prevention of pollution is more cost-effective than its cure. Modern landfill sites are designed with multiple barriers between waste and aquifers (Fig. 16.4a). Clay and plastic liners provide the engineered containment, but low permeability rocks offer a geological barrier. Leachates must be drained and treated, before they overflow the site. Hazardous chemical wastes require purpose-built facilities (Fig. 16.4b). Leakage of **landfill gas**, mostly methane, must be controlled by drilling wells and burning the vented gas, increasingly as a commercially viable fuel.

Sites where land or groundwater have been polluted by their past use present a major legal problem. Attempts to identify and register contaminated land in the UK have so far been frustrated because the threat to the value of such land has taken priority over any health risk to its present or future users.

(a) Contained sanitary landfill

(b) Landfill for hazardous waste

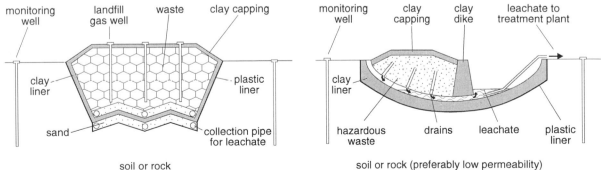

Figure 16.4 Engineering of safe landfills for mixed wastes and for hazardous chemical wastes.

16.3 Diffuse pollution

Overlapping pollution plumes from multiple point sources can contaminate a large aquifer volume. However, some widespread groundwater pollution results from the diffuse nature of its source. Polluted rivers contaminate aquifers that they recharge, particularly their own alluvium. More diffuse is the contamination to percolating soil water from additives to farmland. The most dispersed source is rainwater containing dissolved acid from the atmosphere. It is feasible, if costly, to clean up a local pollution plume in an aquifer, but diffuse pollution must decay naturally by stopping the source of the contamination.

The process of **acid deposition** is a complex interaction of atmosphere, water and soil (Fig. 16.5). Rainwater is naturally slightly acidic due to sulphuric acid (H_2SO_4), mainly derived from the dimethyl sulphide emitted by some marine planktonic organisms. This acidity is increased by the addition of acid-forming chemicals from human activity. The most significant of these are **sulphur dioxide** (SO_2) and the **nitrogen oxides** – nitric oxide (NO) and nitrogen dioxide

(NO_2). In the atmosphere these react first with oxygen to form sulphate and nitrate particles, then with water vapour to form sulphuric and nitric acids. **Dry deposition** is the fallout to the ground surface of the sulphate and nitrate particles, which are later activated by water to give acids. **Wet deposition** is the fallout of atmospheric acids in rain, snow or fog, often after transport of hundreds of kilometres.

The gases that cause acid deposition mostly come from burning fossil fuels (Fig. 16.5). In the UK, total sulphur dioxide emissions have been around 4 million tonnes per year for the past decade, after two decades of falling levels due to the decline of heavy industry. The major contribution is from power stations, particularly those burning coal. By contrast, nitrogen oxide emissions have risen sharply over the past decade from 2.2 to 2.8 million tonnes per year, due to the increase in road transport.

Acid deposition can cause direct damage to plants and to building materials such as limestone, steel and cement. However, most damage to organisms is done as acid waters run off into lakes and infiltrate downwards into soil and rock. An important factor is the

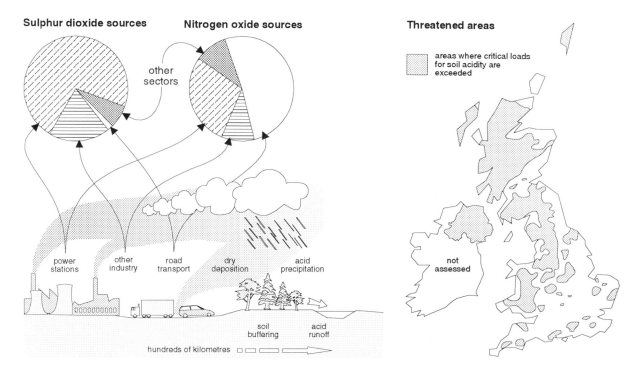

Figure 16.5 Sources of acid deposition and vulnerable areas in the UK.

composition of the soil and its bedrock. Calcareous rocks neutralize the acid water, yielding extra sulphate and nitrate. These compounds are less likely to damage organisms than the residues from siliceous rocks that lack this buffering capacity. Toxic elements such as aluminium, zinc, manganese, copper, cadmium, chromium and lead accumulate in the soil and can weaken plants that absorb them through their roots. The **critical load** of a soil is the amount of acid deposition that it can bear without harm to sensitive elements of its ecosystem. The critical load is currently exceeded by acid deposition over large areas of the UK, particularly on the more siliceous rocks of the north and west (Fig. 16.5).

The other main diffuse cause of polluted groundwater is the infiltration of residues from the agrochemicals used in intensive farming. The contamination from **pesticides** is potentially the more serious, but that from **nitrate fertilizers** is the better studied. Use of these fertilizers has increased by between five and ten times over the past 50 years.

The pollution risk is greatest in an **unconfined** aquifer, lying directly under the soil, or separated from it by permeable rocks (Fig. 16.6a). Measurements in East Anglian Chalk show how concentration peaks for nitrate move down through the unsaturated zone with time, affecting it to depths of some metres (Fig. 6.6b). The risk of contamination is lower where an aquifer is **confined** by less permeable rocks (Fig. 6.6a). These horizons, typically clays or glacial tills, slow the infiltration rate and adsorb pollutants onto the surfaces of their constituent clay minerals. **Denitrifying bacteria** in some groundwaters can convert nitrates to nitrogen, which is released as a gas.

All unconfined aquifers under farmland are at risk from agrochemical pollution. However, the most vulnerable major aquifers in the British Isles are in the area of eastern England where rainfall is low and therefore arable farming predominates (Fig 16.7).

Remedies to the agrochemical pollution of groundwater lie in better matching their application to their uptake by crops and neutralization by soils. Ensuring a crop cover and avoiding fertilizer application during the winter wet season helps to prevent leaching of nitrate. However reduced total agrochemical inputs are the most direct remedy, coupled with change of land use in sensitive areas of groundwater recharge.

(a) Agricultural pollution risk

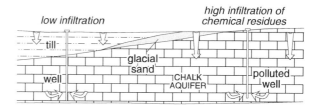

(b) Vertical pollution profile

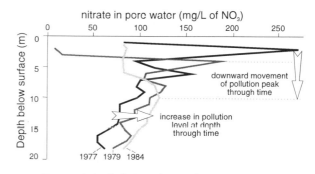

Figure 16.6 Pollution of groundwater by nitrates.

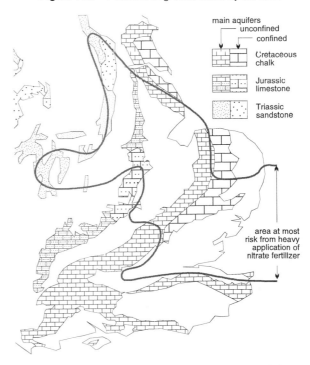

Figure 16.7 Major aquifers at risk from nitrate pollution.

16.4   Radioactive wastes

Radioactive wastes present a special problem because they must be isolated for long periods – thousands or even millions of years. In Britain, weakly radioactive wastes are being buried in lined landfills, mainly near the Sellafield nuclear installation in northwest England. More highly radioactive wastes are being stored at nuclear sites. One option for society is to store such wastes indefinitely where they can be contained and monitored. An alternative option, seen as freeing future generations of this management burden, is to bury wastes in sealed underground repositories.

This disposal strategy is to contain the waste by multiple barriers – metal drums overpacked in concrete, then stacked in a concrete vault later backfilled with clay. The host rock for the vault provides a **geological barrier** when the other barriers leak. Suitable rock bodies slow the migration of radionuclides either by their low permeability to fluids or by their capacity to absorb pollutants. Host rocks for highly radioactive wastes must also maintain a barrier despite heat emission from the waste.

Present UK policy is to cool high-level radioactive wastes in surface storage for at least 50 years before solidifying them for disposal. Attention has therefore focused on the disposal of intermediate- and low-level wastes. Favoured geological settings have low permeability rocks in conditions of stagnant or slow groundwater flow (Fig. 16.8a). Small island sites may have such stagnant saline water below an isolated freshwater lens. Groundwater in igneous or metamorphic rocks is thought to flow mainly through fractures that bound potentially stagnant sites. Any sedimentary cover will tend to constrain most of the water flow in its permeable units. Sites where flow is down towards the centre of a sedimentary basin would ensure a long flow-path for released pollutants.

These geological settings translate into the potential British areas for nuclear waste disposal (Fig. 16.8b). Present assessment efforts are concentrating on a site close to the Sellafield installation. Originally designated for its sedimentary cover rocks, the target is now the underlying fractured volcanic basement. Political and logistic factors have overcome geological reasoning against this problematic site (Ch. 18).

(a) Potential hydrogeological environments   **(b) Potential repository areas in Britain**

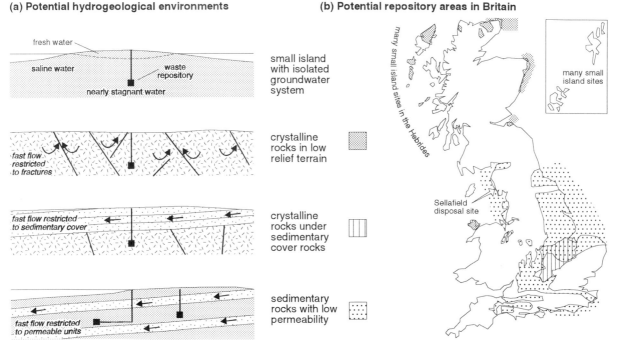

Figure 16.8   Potential geological settings and areas for subsurface nuclear waste disposal in Britain.

16.5   Global perspectives

The pattern of waste management in the UK and Ireland is mirrored in other developed nations. Among these, the wealthy countries tend to generate the most waste, at least from municipal sources (Fig. 16.9). The amount of available landfill space is also an influence. Sparsely populated nations such as the USA, Canada, Australia and New Zealand produce more waste than average, in contrast to crowded countries such as Switzerland, Japan and Luxembourg. Most managed waste is disposed of in the country that produces it. Pollution from bad waste management is therefore often seen as merely a national problem. However, there are several important ways in which wastes are being exported across national boundaries.

Waste disposal at sea is still practised in domestic and international waters. Materials such as dredgings may cause only local physical disruption, but wastes such as chemicals, sewage sludge and ash can perturb chemical and biological systems over wide areas and across national limits. International agreements such as the London Dumping Convention are attempting to regulate waste disposal to the global seas.

A second route for transnational pollution is the waste exported for treatment and disposal. Tightening environmental controls in the developed world make it less costly to process hazardous wastes in countries with low pollution standards. The main exporters are the OECD countries, the importers mostly in Eastern Europe, the Far East and Latin America. Many of these wastes are treated inadequately or not at all, endangering local populations. As with sea dumping, international agreements are now attempting to regulate this transfrontier shipment of waste.

The third source of transnational pollution is from unmanaged or improperly managed wastes. Rivers efficiently transport effluents between countries – over 40% of the world's population live in river basins shared by two or more countries. A more effective dispersing medium still is the atmosphere. For example, Britain exports a proportion of its sulphur dioxide emissions, some of it to Sweden which is reckoned to import nearly 90% of its sulphur-derived acid deposition. Most harmful of all is the dispersal of greenhouse gases, a major topic of the next chapter (Ch. 17).

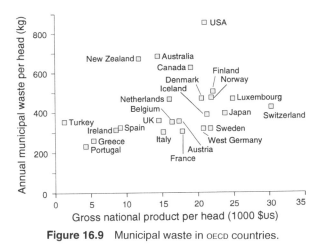

Figure 16.9   Municipal waste in OECD countries.

Further reading

Allen, R. 1992. *Waste not, want not: the production and dumping of toxic waste.* London: Earthscan. [A social and political perspective.]

Chapman, N. A. & I. G. McKinley 1987. *The geological disposal of nuclear waste.* Chichester: John Wiley. [Authoritative but readable treatment of the science and technology of nuclear waste disposal.]

Department of the Environment 1992. *The UK environment.* London: HMSO. [Data on acid deposition (Ch. 2), agricultural pollution (Chs 4, 5), freshwater pollution (Ch. 7), waste disposal (Ch. 11) and radioactive waste (Ch. 13).]

Gourlay, K. A. 1992. *World of waste: dilemmas of industrial development.* London: Zed Books. [A stimulating overview of the production and disposal of wastes.]

Mason, B. J. 1992. *Acid rain: its causes and its effects on inland waters.* Oxford: Oxford University Press. [Authoritative summary of recent research on the impact of acid rain in the British Isles.]

Mellanby, K. 1992. *Waste and pollution: the problem for Britain.* London: HarperCollins. [Review of all forms of waste and pollution in Britain.]

McCormick, J. 1989. *Acid Earth: the global threat of acid pollution.* London: Earthscan. [Popular account of the global impact of acid deposition.]

Park, C. C. 1987. *Acid rain: rhetoric and reality.* London: Routledge. [A readable overview of scientific and political aspects.]

Reid, E. 1990. *Rock solid: the geology of nuclear waste disposal.* Glasgow: Tarragon Press. [Covers the science and policy of nuclear waste disposal in non-specialist language.]

Chapter Seventeen

ATMOSPHERIC POLLUTION

17.1 Local air pollution

At first sight, pollution of the atmosphere does not seem to be geologically relevant. However, there are two ways in which geologists are now involved in this environmental issue. First, many air pollutants are derived from the use of geological resources, particularly from burning fossil fuels. Secondly, these pollutants disturb natural global systems that link the atmosphere with the hydrosphere and biosphere, and with the geological processes of the lithosphere. The extent to which geologists should be concerned by the downstream effects of the resources that they find is an ethical issue, discussed later (Part V). However, this aside, the contribution of Earth history to the prediction of future climate change is reason enough to review atmospheric pollution.

Most potential pollutants in the atmosphere are naturally occurring **gaseous** or **particulate** substances, such as carbon dioxide, sulphur dioxide or smoke. They become an environmental threat in two circumstances – when they are emitted and trapped in locally toxic concentrations, or when they are produced globally at rates that natural systems cannot absorb.

Localized air pollution occurs close to emission sources such as power stations, industrial plants, roads and urban areas in general. However, serious pollution incidents are commonly due to meteorological and topographic conditions that prevent dispersal of pollutants (Fig. 17.1). These **thermal inversions** occur where cold air is trapped below warmer air in a valley or against a range of hills. The lower air mass can be chilled initially by the sea or against rapidly cooling ground on a clear evening. Warm pollutant emissions then rise only as far as the **inversion layer**. Here the pollutants absorb solar radiation, enhancing the inverted temperature gradient and preventing the lower air mass from warming up. In this way pollution incidents may persist for days. The worst on record was in London over four days of December 1952, when over 4000 deaths were caused by high levels of smoke and sulphur dioxide.

Five **primary pollutants** cause most local air pollution (Fig. 17.2). Human-derived emissions of these substances come mostly from burning fossil fuels (Fig. 17.2a,b,d). **Carbon monoxide** (CO) is produced when petrol is incompletely burned, and can be toxic if inhaled. **Nitrogen oxides** (NO_x) – nitrogen dioxide

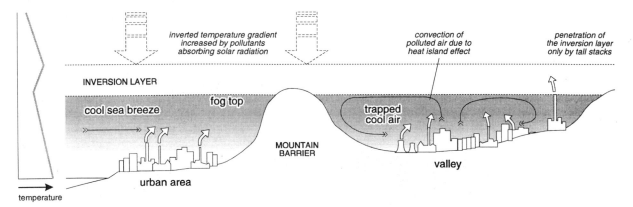

Figure 17.1 Conditions that promote temperature inversions and local air pollution.

(NO_2) and nitric oxide (NO) – form both from nitrogen in fuels and from oxidation of nitrogen in the air. They can retard plant growth, cause breathing difficulties, and are a component of acid rain (Ch. 16.3). Nitrogen oxides react with volatile organic compounds in the presence of sunlight to form ground-level ozone (O_3). This important **secondary pollutant** causes breathing problems and vegetation damage, as well as promoting the reactions that cause acid rain. **Smoke**, mostly carbonaceous particles, is mainly derived from burning coal and diesel fuel. **Volatile organic compounds** (VOC) are produced by evaporation of petrol and vehicle exhausts, together with a large contribution from industrial solvents. **Sulphur dioxide** (SO_2) is derived from coal burning, together with such

industrial processes as oil refining and cement making. It is a major component of acid rain and has a corrosive effect on building materials, plants and animal lungs.

Emissions of SO_2 and smoke in the UK have been falling (Fig. 17.2c) due to decreasing industrial and domestic use of coal. By contrast, emissions of NO_x, VOC and CO have been rising due to the rapid increase in road traffic. Total emissions from this sector have doubled in the past 20 years. The use of emission control devices such as **catalytic converters** is expected to cut local air pollution. However, these emissions contain proportionally more **carbon dioxide** gas, which is an even more serious pollutant of the global atmosphere (Ch. 17.2).

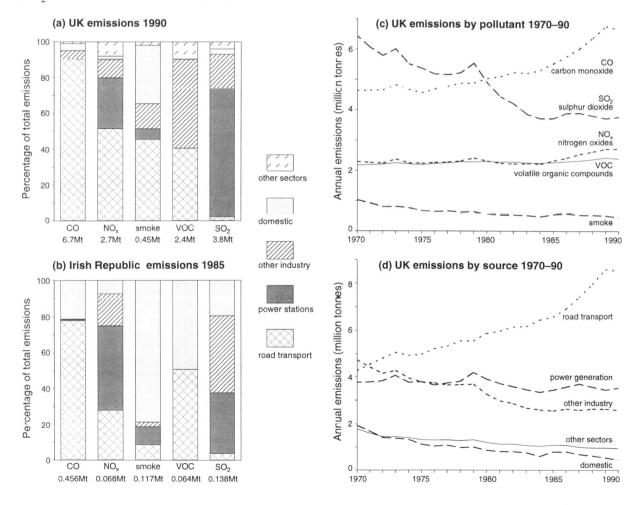

Figure 17.2 Human-derived sources of the gases that cause local air pollution in the UK and Ireland.

17.2 The greenhouse effect

Until the past two decades, air pollution was seen as an essentially local problem. The 5 million billion tonne mass of the atmosphere was assumed to be large enough to disperse human-derived additions and dilute them to harmless concentrations. This assumption is incorrect. A rapid increase in the global concentration of human-derived gases is threatening environmental impacts far from the source of the gas emissions. The most serious threats come not from direct inhalation or contact with these extra gases, but rather from their interference with the radiation budget of the Earth.

The **greenhouse effect** is the insulating tendency of the atmosphere that keeps the Earth's surface warm.

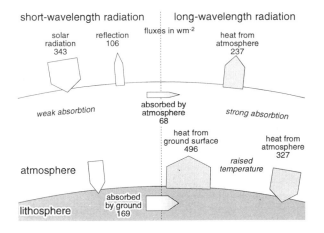

Figure 17.3   Principle of the greenhouse effect.

The insulation is due to the greater transparency of the atmosphere to incoming solar radiation than to outgoing heat (Fig. 17.3). Most atmospheric gas molecules absorb less of the short-wavelength radiation emitted by the hot Sun than the long-wavelength radiation re-emitted by the much cooler Earth.

Of the average 343 watts of incoming solar power that fall on each square metre at the top of the atmosphere, about a third is reflected back to space. Of the rest, a quarter is absorbed by the atmosphere but most is absorbed at the Earth's surface. This radiation heats the ground and is re-emitted with a longer wavelength. However, most of this upward-radiating heat is re-absorbed by the atmosphere and radiated again, downwards as well as upwards. The same amount of outgoing energy is eventually emitted from the top of the atmosphere as the amount of solar energy originally absorbed. However, the recirculation of heat between the lower atmosphere and the ground keeps the Earth's surface about 33°C warmer than it would be without its atmospheric blanket.

The greenhouse effect is vital in maintaining the environment for life on Earth. The current concern is rather over the **enhanced greenhouse effect** – due to human-derived gas emissions promoting **global warming**. The radiative effect of a gas does not depend simply on its concentration in the atmosphere (Fig. 17.4a). Some gases such as **chlorofluorocarbons** (CFCs) are far more potent, molecule for molecule, than others such as **carbon dioxide**. Conversely, molecules of some gases such as **methane** have only

(a) Global warming potentials of gases

	average concentration (ppmv)	lifetime (years)	direct GWP over 100years	sign of indirect GWP
CO_2	355	120	1	none
CH_4	1.72	10.5	11	positive
N_2O	0.31	132	270	uncertain
CFC11	0.00026	55	3400	negative
CFC12	0.00045	116	7100	negative

(b) Warming contribution from 1990 global emissions

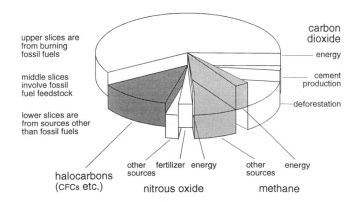

Figure 17.4   Global warming potentials and relative contributions of human-derived gases to the greenhouse effect.

a short **atmospheric lifetime** before they react to form other compounds, in this case carbon dioxide and water. Each gas therefore has a direct **global warming potential** (GWP) over a future timespan, which depends on its molecular structure, concentration and lifetime. So 1 kg of methane released now will cause 11 times more warming, totalled over the next hundred years, than will 1 kg of carbon dioxide (Fig. 17.4a). Some gases also have an indirect GWP due to the potential of their reaction products.

Although **water** is the most potent natural **greenhouse gas**, the most threatening human additions are **carbon dioxide** (CO_2), **methane** (CH_4), **nitrous oxide** (N_2O) and **halocarbons** (CFCs and related gases) (Fig. 17.4b). As with the causes of local air pollution, it is notable that most greenhouse gases derive from geological resources – either directly from burning fossil fuels or indirectly from using their manufactured products. Predominant is the carbon dioxide from burning oil, gas and coal.

The measured average concentrations of human-derived greenhouse gases have risen sharply over the past 40 years (Fig. 17.5). The concentration of nitrous oxide is now 8% above that before the industrial revolution, carbon dioxide is 27% higher and methane 115%. CFCs are a novel addition to the atmosphere since the 1930s. Human-derived gases therefore form a substantial new component of the atmospheric heat blanket. Over the past 100 years of detailed measurements, the global average surface temperature has risen by between 0.3°C and 0.6°C (Fig. 17.6). The magnitude of this rise is consistent with computer models of global warming over the same period, but is nevertheless still within the range of natural climate variability. Another decade of data will be needed to show unequivocally the human-induced warming.

The same computer models predict warming rates of between 0.2°C and 0.5°C per decade over the next century (Fig. 17.6). The uncertainties in these predictions arise from the complexity of atmospheric processes, particularly cloud formation, and of the interaction of the atmosphere with the oceans and biosphere. Prediction also involves choosing an appropriate **emissions scenario** – a pattern of greenhouse gas production over the next century. These scenarios are strongly dependent on extrapolating past patterns of fossil fuel use into the future (Ch. 12).

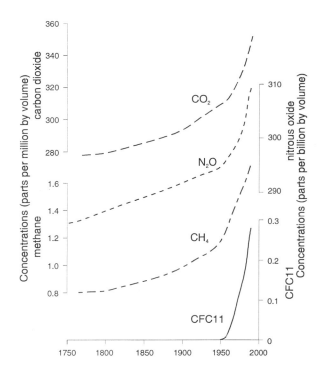

Figure 17.5　Rise in global levels of greenhouse gases.

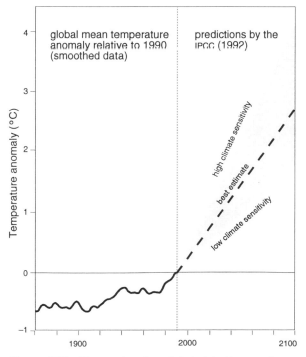

Figure 17.6　Observed and predicted global temperatures.

17.3 Geology and global warming

The build-up of greenhouse gases is controlled not just by atmospheric processes, but by **Earth systems** that involve the biosphere, hydrosphere and also the lithosphere. The **carbon cycle** is the most important example (Fig. 17.7). Carbon is held in **reservoirs** on the Earth, for instance as carbon dioxide in the atmosphere, as inorganic carbonates dissolved in the oceans, as living tissue in the biota and as coal and hydrocarbons in fossil fuels. **Fluxes** of carbon between these reservoirs occur mainly by natural processes – **photosynthesis** by plants, **respiration** by animals, **decomposition** of organic matter, and **evaporation** from or **rainout** back to the oceans. The fluxes through each reservoir nearly balance on a human timescale. This balance is being disturbed by two human-derived fluxes to the atmosphere – from the land biota by **deforestation** and from the fossil fuel reservoir by **combustion**. Although these additions, now about 7.5 Gt of carbon each year, are less than natural fluxes, they are proving too large a burden for the atmosphere. About 3.5 Gt of carbon remain there each year, with only 3 Gt being removed elsewhere, mainly to the oceans.

There is a geological imbalance between the 6 Gt taken each year from the crustal carbon reservoir and the far slower rate of formation of carbon-bearing rocks. The eventual depletion of fossil fuels has already been discussed (Ch. 12). However, the arithmetic of the carbon cycle shows up a more urgent problem of sustainability – the inadequacy of the atmosphere to hold the waste carbon from the large fossil fuel reservoir. Burning only the 1500 Gt in proved reserves of conventional fuels would double the present carbon dioxide level of the atmosphere, even assuming that natural processes kept removing about half the extra carbon to other reservoirs.

The geological process of carbon removal is by **sedimentation**, mainly from the oceans. Some carbon is bound in **carbonates** by the organisms that form calcareous hard parts. The rest accumulates as dead soft tissue in **organic-rich sediments**, mostly muds. One hope is that the carbon flux by sedimentation will increase in response to high oceanic levels of dissolved carbon dioxide, locking away more excess carbon. There is no evidence that this is happening. Indeed, factors such as ocean pollution and warming, and increased levels of ultraviolet radiation (Ch. 17.4), threaten to hamper the ability of organisms to pump down carbon in this way. Rocks potentially provide an important record of historical fluctuations of organic productivity to compare with future trends.

Volcanism is another natural flux of carbon from the crust to the atmosphere. However, volcanoes emit less than 1% of the carbon dioxide produced by human

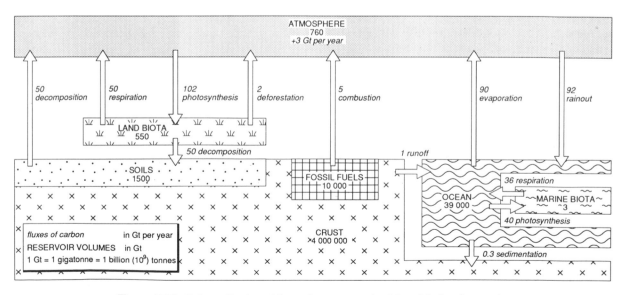

Figure 17.7 Schematic view of the main components of the global carbon cycle.

(a) Distribution of known methane clathrates

(b) Carbon in clathrates compared with other fossil fuels

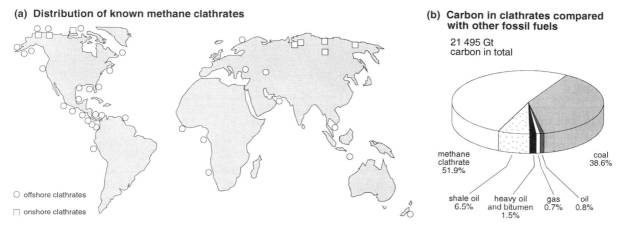

21 495 Gt carbon in total

methane clathrate 51.9%

coal 38.6%

shale oil 6.5%

heavy oil and bitumen 1.5%

gas 0.7%

oil 0.8%

○ offshore clathrates
□ onshore clathrates

Figure 17.8 Known occurrences of methane clathrates and their estimated total global abundance.

activity. A far larger threat is that from **methane clathrates**, the frozen methane locked in permafrost and beneath continental margins (Ch. 11.7; Fig. 17.8a). The carbon content of clathrates probably equals that in the rest of the fossil fuel reservoir (Fig. 17.8b). Burning any of this methane would add to the atmospheric CO_2 level. However, the release of clathrates as methane gas would be far more serious, because the global warming potential of CH_4 is greater than that of CO_2 – 11 times more over 100 years, and 35 times over 20 years. The tundra areas of Asia and Canada, where onland clathrates are most common, are ominously the areas where the largest local temperature rises are now being recorded. Geological surveys need to define better the location and extent of the methane clathrates at most risk.

The most distinctive contribution of geology to assessing global warming comes from charting past patterns of global change, as a guide to future trends. An example is the estimate of sea level change over the past 160 000 years from coastal sediments in New Guinea (Fig. 17.9d). This curve closely matches data on levels of greenhouse gases trapped in bubbles in Antarctic ice of the same age, and on local temperature derived from oxygen isotopes in the same ice (Fig. 17.9a, b, c). Such correlations are powerful warnings of potential global impacts, even if some details of cause and effect are unclear. The warning is strong, in that present-day levels of carbon dioxide and methane (Fig. 17.5) are already off the scale of this recent geological record (Fig. 17.9).

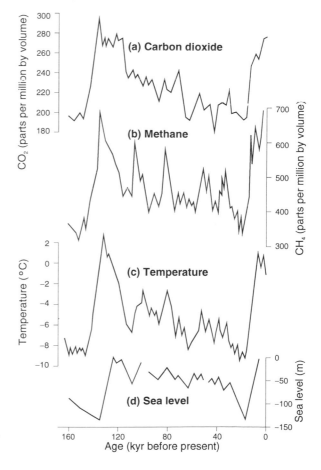

Figure 17.9 Atmospheric composition and temperature measured in the Vostok (Antarctica) ice core, compared with global sea level estimated from New Guinea.

17.4 Stratospheric ozone depletion

Ground-level ozone is a harmful pollutant (Ch. 17.1), but ozone in the lower stratosphere acts as a vital absorbent of harmful ultraviolet radiation. Ozone (O_3) and oxygen (O_2) molecules reach a dynamic balance with each other as they react in sunlight, and normally maintain a rich **ozone layer** between about 15 and 25 km. However some catalysts at this level favour the reaction of ozone to oxygen, lowering the ozone concentrations.

Ice crystals are a natural catalyst during polar winters, particularly in the Antarctic where circum-polar winds isolate the cold polar air. By the Antarctic springtime a depleted **ozone hole** develops, surrounded by ozone rich air of the circumpolar vortex (Fig. 17.10a). This effect is amplified by human-derived catalysts, particularly chlorine atoms from the dissociation of **halocarbon** molecules. The dramatic rise in the use of these compounds, particularly CFCs (Fig. 17.5), correlates with a sharp fall in the effective amount of ozone in the Antarctic hole (Fig. 17.10b).

Chlorine emitted from volcanoes as hydrochloric acid is rained out in the lower atmosphere, and does not add to the stratospheric chlorine load. However, the dust and sulphuric acid aerosols injected into the stratosphere by very powerful explosive eruptions do promote ozone depletion, by providing nucleii for reactions involving human-derived chlorine atoms. Eruptions such as that of Mount Pinatubo in 1991 can therefore trigger enhanced ozone depletion, in low latitudes as well as over the Poles.

Further reading

Bailey, M. C., J. J. Bowman, C. O'Donnell 1986. *Air quality in Ireland: the present position.* Dublin: An Foras Forbatha. [The most recent survey of local air pollution in the Irish Republic.]

Department of the Environment 1992. *The UK environment.* London: HMSO. [Chapter 2 reviews local air quality and Chapter 3 the UK contribution to global atmospheric pollution.]

Elsom, D. K. 1992. *Atmospheric pollution: a global problem.* Oxford: Basil Blackwell. [An introduction to local and global pollution of the atmosphere for the non-specialist.]

Houghton, J. T., B. A. Callander, S. K. Varney (eds) 1992. *Climate change 1992: the supplementary report to the IPCC scientific assessment.* Cambridge: Cambridge University Press. [An essential update to the first report cited below.]

Houghton, J. T., G. J. Jenkins, J. J. Ephraums (eds) 1990. *Climate change: the IPCC scientific assessment.* Cambridge: Cambridge University Press. [The first consensual assessment of the global warming risk by climate scientists.]

Gribbin, J. 1990. *Hothouse Earth.* London: Bantam. [A popular introduction to global warming.]

Leggett, J. (ed.) 1990. *Global warming: the Greenpeace report.* Oxford: Oxford University Press. [A commentary on the first report of the UN IPCC, but an excellent summary of the subject in its own right.]

Nisbet, E. G. 1991. *Leaving Eden: to protect and manage the Earth.* Cambridge: Cambridge University Press. [A readable summary of the complexities of atmospheric pollution in the wider context of global change.]

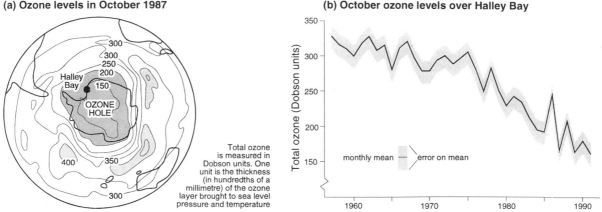

(a) Ozone levels in October 1987

Halley Bay

OZONE HOLE

Total ozone is measured in Dobson units. One unit is the thickness (in hundredths of a millimetre) of the ozone layer brought to sea level pressure and temperature

(b) October ozone levels over Halley Bay

Total ozone (Dobson units)

monthly mean — error on mean

Figure 17.10 The Antarctic ozone hole in September 1987, and the trend of ozone depletion over Halley Bay.

V GEOLOGY
AND SOCIETY

Previous parts of this book have surveyed the main components of environmental geology, particularly as they apply to the British Isles. These components have been described as objectively as possible, although many of the case studies highlight subjects of current political and economic controversy. This final part of the book draws out some of these broader issues from the science of environmental geology. How should geology relate to other environmental sciences? Should it be detached from or involved with the decision-making processes of society? How far should ethical concerns about the environment determine the future course of geology?

Emissions from the burning of fossil fuels pose the most serious threat to the environment of any geological resource.

Chapter Eighteen
GEOLOGY AND SOCIETY

18.1 History of applied geology

The term "environmental geology", as used in this book, is relatively recent. It dates from the America of the 1960s, and was popularized there in books such as those by P. T. Flawn (1970), D. R. Coates (1971), R. W. Tank (1973) and F. Betz (1975). However, much of the subject matter of environmental geology has a far longer history, most of it grouped as "applied" or "economic" geology. A brief survey of this history is instructive, revealing the roots of geologists' traditional perception of their rôle within industrial societies. Later parts of this chapter suggest that this perception must change rapidly if geology is best to serve the future interests of society.

Pure and applied aspects of geology have cross-fertilized each other since the early days of geology in Britain. William Smith, who devised the first integrated scheme of British stratigraphy, made many of his observations as an engineer on the rapidly growing canal network of the late eighteenth century. In-

deed, he delayed publication of his scheme for 20 years, until 1816, because he thought that his data had great commercial value. However, the main use of 19th century geology was not to be within engineering, but in finding minerals and coal to fuel industrial expansion.

In the early 19th century, mining relied on empirical methods to exploit accessible, readily identifiable, high-grade deposits. Little scientific skill was required to find these, but as such deposits were progressively exhausted, the value of geological exploration was appreciated. The Geological Survey of the United Kingdom was established in 1835 with the explicit task of mapping the nation's rock formations and evaluating its mineral wealth. Under the directorship of Henry De la Beche, the Survey expanded to include a Museum of Practical Geology and, in 1851, the Royal School of Mines. The enlarged organization widened its geographical brief to include Britain's colonial dependencies, where fresh mineral resources were being sought to substitute for dwindling domestic reserves.

The importance of geology in the development of the British Empire increased further after 1855 under the new director of the Geological Survey, Sir Roderick Murchison. Also president of the Royal Geographical Society, Murchison was a pivotal figure of mid-Victorian imperial exploration. Geology and geography provided much of the information about newly colonized lands that was needed to exploit them efficiently. Although this colonial geology had a primarily economic motive, pure science gained from the rapid importation of fresh data to test and amplify its theories. Scientific empires were built too, through campaigns of overseas field work, the winning of new data, specimens and maps, and the prizes of publications and careers.

The rôle of geology as a provider of the Earth's resources persisted long after Murchison's death in 1871. Indeed, it outlived the British Empire itself to

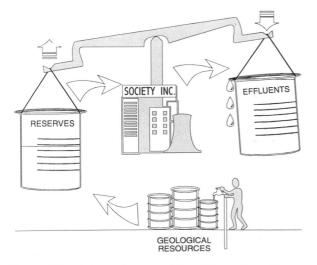

Figure 18.1 A view of the geologist's preoccupation with providing resources rather than monitoring their effluents.

become the principle that guided applied geology into the late 20th century. This principle embodies the **frontier ethos**, in which Earth resources are tacitly regarded as inexhaustible, and merely awaiting discovery and exploitation by any power with sufficient scientific and technical knowledge. The vocabulary of the petroleum industry still includes terms such as "frontier oil province" which perpetuate this exploration philosophy. Most geologists are paid to satisfy the rising demand for natural resources without being encouraged to worry about the impacts of using these resources – in particular where the effluents from them might be going (Fig. 1).

Although the provision of resources has dominated applied geology for most of this century, there has been a shift in concern away from minerals, particularly metallic minerals, towards fossil fuels and water. This changing emphasis is clearly reflected in the content of some representative textbooks on applied geology published in Britain (Fig. 18.2). So, in 1928, J. W. Gregory devoted nearly three-quarters of his *Elements of economic geology* to mineral deposits. This proportion had dropped to 60% by 1944, in Fearnsides & Bulman's *Geology in the service of man*, and to just a quarter of Watson's *Geology and man*. A more recent and fundamental shift in focus has been away from geological resources towards the environmental impacts of human activity, in the wake of the surge in environmental science of the 1970s.

In fact, this second strand of environmental geology is not new. Concerns for impacts on the geological environment and from the use of geological resources

date from when Murchison was at the height of his influence, with G. P. Marsh's *Man and nature: physical geography as modified by human action* (1864). Sporadic later works developed the same theme, most notably in Britain with R. L. Sherlock's *Man as a geological agent* (1922). This prescient book catalogued the whole range of human influences on the geological environment, including a calculation of the magnitude of the global warming risk from burning fossil fuels. However, the thread of geologists' concern for environmental impacts through the first half of this century is so thin that it can be recognized only in retrospect.

Even the more recent growth in anxiety over human impacts on the environment has yet to filter deeply through the geological profession. One index of the continuing focus on geological resources is the field of work in which new geologists have been employed in recent years (Fig. 18.3). In the UK, nearly half have been joining the petroleum industry or the companies that service their exploration effort. Another 12% have entered consultancies, mostly clients of the oil industry. Add to the total those involved in extracting water, coal, minerals and construction materials, and nearly three-quarters of geologists in the UK are probably employed primarily to find resources. This situation is beginning to change, with a demand for advice on impacts to the geological environment, particularly those of groundwater pollution. The geological profession, like society in general, is at a pivot point where concern for the effluents from society is set to outbalance the traditional preoccupation with where its resources are to come from (Fig. 18.1).

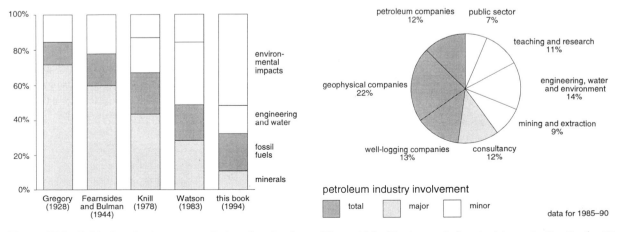

Figure 18.2 Subject matter in some applied geology books.

Figure 18.3 Employment of geologists graduating the the UK.

18.2   Science in society

The recent history of applied geology has seen its scope broaden and its focus shift in response to wider concerns within society. Before examining the nature of these "environmental" issues, it is important to recognize that geology is indeed influenced by, and can in turn influence, the changing climate of world politics, economics and ideas. The alternative view is of geology as a more isolated endeavour, setting its own agenda, and generating objective "scientific" results which then may or may not prove useful to society.

The history of science does not support this second view, most popular among scientists themselves, that a discipline such as geology is unaffected by the contemporary culture in which it operates. The shifting emphasis of applied geology, from metallic minerals and coal towards groundwater, industrial minerals and petroleum, has been a direct response to the changing demands of society for geological advice. The direction of "pure" geological research has, in turn, been altered by the theoretical needs of new types of exploration and by the novel geological data that it generates. So, recent understanding of the formation and filling of sedimentary basins has been stimulated by the quest for the oil and gas accumulations which these basins host. The field of hydrogeology is progressing rapidly because of society's need to bury waste, including radioactive waste, in a way that will not threaten groundwater quality. Even plate tectonics, the geological theory of the century, sprang from the ready availability of government defence spending in the climate of the Cold War. The vital data on earthquake distributions came from instruments installed to monitor nuclear tests, and the better charting of the ocean floors was needed for navigation of submarines.

Even if the historical course of geology has been influenced by external forces, perhaps the theories that result are nevertheless untainted by cultural influence. This is a much-debated subject, with most scientists arguing the purity of their work and most others questioning it. Scientists may certainly aspire to the ideals of "openness, impartiality and self-criticism" in their research, but these ideals are not always reached. With so much geological research now contractual to and funded by industry, it is understandable if hypotheses that seem useful to the clients are favoured over those that do not. Subjectivity in science is increasingly evident where geologists employed on two sides of a public or industrial enquiry give conflicting scientific advice – for instance, over the safety of a proposed underground store for nuclear waste, the source of pollution in groundwater, or the allocation of petroleum reserves between companies in a jointly owned field.

An increasingly heard view is that the worthy ideals of the scientific method would be better protected by recognizing the cultural challenges to them rather than by pretending that these challenges do not exist. Science has always been confronted by and had to respond to the broader ideas and attitudes of the time. The recently developed ideas about the environment are no exception.

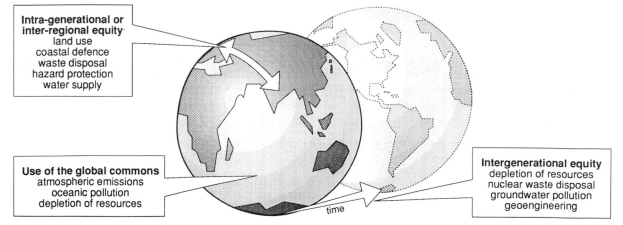

Figure 18.4   Components of environmental ethics, with some geologically related examples.

18.3 Geology and environmental ideas

What are the main strands of environmental thought that are influencing the course of applied geology? How should geologists respond?

A central environmental view, from which many other ideas spring, is of a **finite world**, with some of its systems being stressed to their limits by human activities. This perception is not new, but has grown in parallel with the exponential increases in population, resource use and waste production. The image of "spaceship Earth" from the Apollo moon missions was a particularly vivid challenge to the view of a limitless world that merely awaited further human exploitation. Yet this view still underlies the world's predominant economic systems, in which progress is synonymous with economic growth and its increased throughput of resources and energy.

The prevailing theories of human economics derive from a **utilitarian ethic**, which regards natural resources as available for appropriate human use. The opposite view is the **preservation ethic**, by which human beings have no special rights over the rest of nature, and should leave it as they found it. A less radical response to the view of a finite Earth is the **conservation ethic**, which favours minimizing human impacts so that natural systems are sustainable in the long term. However, such an obviously difficult compromise requires specific ethical principles to guide detailed policies and actions. These emerging principles can be related directly to particular issues in environmental geology (Fig. 18.4).

The principle of **intergenerational equity** is perhaps the most relevant of all to geologists – the idea that the activities of future generations should not be jeopardized by those of present generations. The foremost geological contradiction to this principle is the depletion of subsurface resources. Minerals and fossil fuels are all being used many times faster than they are regenerated by geological processes. Even where the volume of a resource is large, such as that of coal, future generations will be forced to exploit lower-grade and less accessible deposits, with ever-increasing energy costs. The geologists who find these deposits may increasingly be involved in conserving them, maximizing the social benefit rather than the monetary profit from their use.

Intergenerational equity is also threatened by poor conservation of groundwater. This resource is sustainable only if it is protected from contamination and is extracted no faster than it is replenished. Whereas polluted sources of surface water can often be cleaned within years or decades, degraded groundwater may persist for centuries. Hydrogeologists are now having to emphasize that the prevention of groundwater pollution is far easier than its cure. The proposed underground disposal of nuclear waste is a test case of special significance, because long-lived radionuclides present such a clear intergenerational risk.

A second principle of environmental ethics is that of **inter-regional equity** (or intra-generational equity) – the axiom that no one section of human society should suffer unreasonably from the activities of others. A groundwater supply in one area may be polluted by effluents discharged elsewhere into the same aquifer. A protection scheme for one stretch of coastline may increase erosion on another stretch. A motorway built to ease traffic problems in southeast England may mean expanding quarries elsewhere in the country. These are local geological issues of inter-regional equity. More serious equity conflicts occur on a global scale. Populations in undeveloped regions are disrupted to open mines supplying metal ore to the developed world. Toxic waste is shipped from countries with tight environmental controls to less regulated countries equally at risk from pollution. Over such issues, society needs geologists to take a broad view of the comparative impacts, rather than to focus only on the benefits for a client in one area.

A special aspect of inter-regional equity is the fair **use of the global commons** – the components of the Earth, such as the atmosphere and oceans, which are regarded as beyond national ownership. A prime example of abuse of the commons is the excessive burning of fossil fuels by a small proportion of the world's population, emitting greenhouse gases that degrade the atmosphere for the use of the majority (Ch. 17). How geologists respond to such abuses of the commons will depend on their view of **downstream liability** – the concept that geologists bear some responsibility for the impacts of using the commodities that they find. Should resources such as fossil fuels themselves be included in the global commons, and managed by international agreement?

18.4 Geology and the future

This book has surveyed the components of environmental geology from a European perspective in the 1990s. The balance between these components would be different if viewed from another part of the world. Moreover, the emphasis of "applied geology" will be different again in the next decade, as the geologists respond to the shifting concerns of society. How will, or indeed should, the subject develop in the future?

The need for geologists to focus on **environmental impacts** as well as on supplying **resources** has already been portrayed (Fig. 18.1). Barriers to this widening of concern are provided by the divisions between traditional sciences. Environmental impacts rarely affect only the subsurface, nor do they derive only from using geological resources. The recognition, monitoring, treatment and prevention of a threat such as acid deposition (Ch. 16.3) needs collaboration between geologists, soil scientists, biologists, hydrologists and atmospheric scientists. Most geologists trained in the British Isles have so specialized an education that they may be unable to appreciate the full nature of the threat, or to communicate effectively with other scientists towards its understanding.

All sciences have become more **reductionist**, dissecting natural systems into their smallest components to study basic processes. However, some natural systems can be understood only by re-integrating these processes at a larger scale, often across traditional scientific boundaries. Most environmental problems need this **holistic** strategy. **Earth systems science** is one response, emphasizing the interconnectedness of natural processes on all scales. One of its analytical devices has been used in this book – the concept of a **physical** or **biogeochemical cycle** that shows the flow of matter or energy through natural systems. Examples are the carbon cycle (Fig. 17.7) and the hydrological cycle (Fig. 7.1). Such representations provide frameworks on which to hang the results of more detailed research, allowing scientific specialists to see the wider significance of their results.

Environmental geology will increasingly need the systems approach both to teach the subject and to collaborate with other scientists in solving environmental problems. A growing educational dilemma is whether to teach holistic **environmental science** with-

out specialization, or to teach it as a supplement to traditional subjects such as geology. Society probably needs the products of both approaches – generalists who know where to find relevant detailed research, and specialists with an environmental awareness.

Geology is a good vehicle for the holistic approach to science. Indeed, geology *is* the systems science of the past Earth. Interpretations of ancient environments preserved in the rock record need the full integration of physical, biological and chemical evidence. Geologists are already used to extrapolating the natural cycles of the present Earth back into the past. They need to grow more confident in reversing this strategy, using the past as a key to present environmental change and to those future changes threatened by human impacts on natural systems.

The systems approach is relevant to local problems and to the Earth as a whole. Geologists, for instance, are familiar with **plate tectonics** as a mechanism for cycling matter and energy on a global scale. Even more general Earth models are possible. James Lovelock's **Gaia hypothesis**, for instance, treats the Earth as a complex self-regulated system. Instead of life being passively dependent on a separate physical and chemical habitat, the Gaia hypothesis sees the biosphere as a key component in forming and maintaining local and global environments. Such models highlight the vital question of whether the Earth's natural life-support systems are now robust enough to regulate the excesses of the human species. Even more important, global systems models provide a rigorous "scientific" methodology for finding answers. The geological record itself preserves the evidence of past fluctuations in local and global environments against which present threats and future predictions can be measured. Only with a feel for the length of geological time can the speed and severity of present human impacts be fully appreciated.

The systems approach may also help to integrate scientific thinking with that of disciplines such as economics, politics and sociology. The cycles of systems science can be seen as **natural economies**, and directly comparable with the **human economies** normally outside scientific concern. Such comparisons reveal striking contrasts which are at the heart of current environmental problems. In natural economies, such as the hydrological cycle (Fig. 7.1), the budgets always balance. However, many human economic systems, such as the petroleum "cycle" (Fig. 13.3) are essentially linear. Resources are extracted from one natural reservoir and effluents are pumped out into others on the assumption that all reservoirs are infinitely large. The faster this "cycle" goes, the healthier the human economy is regarded, ignoring its unsustainable nature and its wider environmental costs. Geologists must join other environmental scientists in the criticism of such human systems, especially where their logic seems counter to that of natural economies.

So, progress in tackling environmental problems involves scientific collaboration across the traditional subject divides. Even more important, it involves breaking down the barriers that still separate science from the rest of human activity. The purely scientific answers to many environmental problems are already known, at least well enough to recommend a prudent course of action. Cutting the use of agrochemicals is the best way of safeguarding many future groundwater supplies. The wise way of reducing the risk of global warming is to burn far less fossil fuel. Yet scientists have been unable to convey to decision-makers and to society in general the severity and urgency of some of these environmental threats. The responsibility for this poor communication lies in each of the "two cultures" of science and the humanities. Public understanding of science is often too poor to assess the scientific message, but many scientists lack the motivation to explain their results to a wider audience. Scientists too often see their rôle as advisers, operating in a **reactive** mode to the questions that society asks. However, until general scientific literacy is higher, scientists have a duty to be more **proactive**, alerting society as lucidly as possible to identified environmental threats.

There is further rôle for environmental geology – that of education, to widen the understanding of science in society. For most people "the environment" exists at and above the Earth's surface – forests and rivers, the oceans and the atmosphere. It extends neither below the ground nor back through geological time. This book is one attempt to broaden the professional view of geologists about the environment and their rôle in it, and to show the more general reader the distinctive environmental perspective that geology has to offer.

Further reading

Coates, D. R. 1981. *Environmental geology*. New York: John Wiley. [Chapters 1–3 contain a helpful discussion of ethical and philosophical problems behind environmental geology.]

Cooper, D. E. & J. A. Palmer (eds) 1992. *The environment in question: ethics and global issues*. London: Routledge. [A stimulating collection of articles that review current themes in environmental ethics.]

Knill, J. L. (ed.) 1978. *Industrial geology*. Oxford: Oxford University Press. [A survey of applied geology in Britain that reflects the beginnings of environmental concerns.]

Lovelock, J. 1988. *The ages of Gaia: a biography of our living Earth*. Oxford: Oxford University Press. [The most geologically orientated of books on the Gaia hypothesis.]

McKibbin, B. 1990. *The end of nature*. New York: Viking Penguin. [A stimulating view of the necessity to manage the Earth.]

National Research Council 1993. *Solid-Earth sciences and society*. Washington: National Academy Press. [A well illustrated manifesto for the systems science approach to geology in the USA.]

Stafford, R. A. 1989. *Scientist of empire: Sir Roderick Murchison, scientific exploration and Victorian imperialism*. Cambridge: Cambridge University Press. [An authoritative account that reveals the rôle of geology in Victorian society.]

United Nations 1993. *The global partnership for environment and development: a guide to agenda 21*. New York: United Nations. [Outlines the Rio declaration from the 1992 Earth Summit. Includes the 27 principles – one formulation of a set of global environmental ethics.]

Watson, J. 1983. *Geology and man: an introduction to applied earth science*. London: Allen & Unwin. [A clearly written introduction in a British perspective, with a growing awareness of environmental impacts.]

The following four books were mentioned in this chapter mainly for their historical interest. However, Sherlock's book in particular makes interesting reading as an early statement of concern over human impacts on the geological environment.

Fearnsides, W. G. & O. M. B. Bulman 1944. *Geology in the service of man*. Harmondsworth: Penguin.

Gregory, J. W. 1928. *The elements of economic geology*. London: Methuen.

Marsh, G. P. 1864. *Man and nature: physical geography as modified by human action*. New York: Charles Scribners.

Sherlock, R. L. 1922. *Man as a geological agent*. London: Witherby.

GLOSSARY

The glossary defines terms which may be unclear to readers without a geological training, and words from the wider field of environmental studies which may be unfamiliar to geologists. The terms are selected from those shown in bold face in the text or in the labels to figures. The glossary does not repeat terms that appear in only one place in the book and are adequately defined there. These terms all appear in the index.

Acadian tectonic event or orogeny of late Silurian or Devonian age also termed late Caledonian

acid deposition fallout of suspended or dissolved acidic material from the atmosphere, either as dry dust or in rain, snow or fog

aggregate particulate material of sand size and greater used in the construction industry – may be naturally occurring or mechanically crushed rock

airfall volcanic material deposited from an airborne cloud rather than from a lateral flow over the ground surface

andesite a volcanic rock with a proportion of silica intermediate between that of basalt and rhyolite

anticline fold with the oldest rocks in the core

artesian well a well or borehole in which water rises to the ground surface under its own pressure

aquiclude rock body which retards or prevents the flow of underground water

aquifer rock body with enough pore space to store underground water and enough interconnection of the pores to permit flow of water through them

asphalt either the naturally occurring solid phase of petroleum or, in engineering usage, a mixture of bitumen and aggregate used for road surfacing

asthenosphere zone of the Earth's interior deeper than about 100 km

avalanche rapid downslope movement of snow, rock or soil

basalt fine-grained igneous rock with a low percentage of silica

base flow the rate of throughput of water in a river during fair weather, rather than during floods

basement older, harder rocks below a cover of younger softer rocks

basin inversion contraction, uplift and deformation of a previously subsiding sedimentary basin

bauxite rock composed of hydrated aluminium oxides, a major source of aluminium ore

bedding depositional surfaces in sedimentary rocks, typically near to horizontal at the time of formation

bimodal having two maxima on a distribution curve of a natural parameter such as grain size or mineral abundance

biogeochemical cycle circulation of a chemical element or compound through organic and inorganic systems

biosphere zone near the Earth's surface that includes all living organisms

bitumen heaviest fraction of refined or natural petroleum

boulder clay or **till** glacial deposit with coarse material supported in a fine grained matrix

brittle deforming by slip on discrete fractures rather than by pervasive flow

buffering chemical or physical absorption of a pollutant, for instance by reaction between groundwater and soil or rock.

Caledonian applied generally to deformation events of Ordovician, Silurian or Devonian age

calorific value the amount of heat energy available from a specific mass of fuel

caprock or **seal** impermeable layer preventing the upward escape of oil or gas from a reservoir

capillary action rise of water through narrow pores, promoted by surface tension

cementation partial or complete filling of pore spaces in sedimentary rocks by crystallization of mineral matter

CFCs chlorofluorocarbons – a family of man-made gases, implicated in the enhanced greenhouse effect and destruction of the ozone layer

chalk very fine-grained, poorly cemented limestone, restricted to Upper Cretaceous and Palaeogene sequences in the British Isles and region

chilled margin rapidly cooled, fine-grained outer zone of an igneous intrusion

china clay rock containing abundant kaolinite, derived by weathering of granite

cirque spoon-shaped erosive hollow formed by ice

clastic composed of discrete sedimentary particles or clasts

clathrate a solidified mixture of gas in ice, usually applied to frozen methane hydrate

cleavage   pervasive planar structure formed by rock deformation, particularly by the rotation and growth of mineral grains

cohesion   strength of a rock due to cementation or electrostatic attraction across potential failure planes

compaction   reduction in volume of a rock unit by physical rearrangement of grains

compressive strength   stress at which a rock fails under a force exerted normal to its surface

concentrate-and-contain   pollutant management strategy involving localized confinement rather than dispersal

confined aquifer   aquifer overlain by impermeable rocks

cover   younger, softer rocks over a basement of older, harder rocks

crude oil   unrefined oil extracted from the subsurface

crystalline   rock composed of minerals grown in place rather than of transported particles – applied to igneous and metamorphic rocks and to evaporites

cultural environment   ambience of human ideas and activities not included in the natural or built environments

degraded resources   natural resources, particularly land and water, whose quality has suffered by exploitation, so that they are less suitable for future use

derelict land   land abandoned from a previous use which has degraded it for most other purposes

desalination   production of fresh water by evaporation and condensation of sea water

dilute-and-disperse   pollutant management strategy involving release into a large body of water or air

dimension stone   stone that can be cut to size and shape for construction purposes

dissemination   mineral deposit comprising ore dispersed through the host rock rather than concentrated in a vein or pod

dolomite   a mineral form of calcium magnesium carbonate

downstream   towards the later stages of an industrial or economic process

drift   Quaternary sediments on the present land surface

dry gas   natural gas that does not yield a liquid phase on cooling

ecosphere   the biosphere and those parts of the atmosphere, hydrosphere and lithosphere with which organisms interact

effective rainfall   difference between the amount of rainfall on an area and the amount of evaporation from it

emissions scenario   an estimate of future pollutant production over time, for instance of greenhouse gases

epigenetic mineralization   formation of a mineral deposit later than the formation of its associated rocks

erosion   breakdown of rock involving transport of material by water, wind, ice or gravitational force

evaporite   sedimentary rock formed by crystallization from a saline solution, usually due to evaporation from a lake or sea

fault   fracture in rock across which there has been shear displacement

fen   bog formed in a valley or estuary, in or next to a permanent body of water

FGD   flue gas desulphurization – absorbtion of sulphur dioxide by reaction with powdered limestone

fissility   property of some sedimentary rocks to split into thin sheets, due to alignment of minerals during compaction

floodplain   the nearly horizontal bottom of a river valley over which water may spill during floods

freeze/thaw weathering   fragmentation of rock by repeated freezing and expansion of water in cracks

fuller's earth   montmorillonite clay, named after its former use for fulling – the cleansing and conditioning of cloth

Gaia hypothesis   the idea of the Earth as a self-regulating natural system

gangue   non-economic minerals in or associated with an ore deposit

gas hydrate   a solidified mixture of gas in ice, usually applied to frozen methane hydrate or clathrate

gas-prone source   hydrocarbon source rock with organic components that produce more gas than oil on burial

geomedicine   relationship between the geology of an area and the health of its humans or animals

geomorphology   study of the shape and origin of landforms

geothermal energy   energy extracted from hot subsurface rocks or water

global commons   resources such as the atmosphere and oceans which are beyond national ownership

gneiss   metamorphic rock with a foliation defined by coarse compositional banding

grade of metamorphism   degree of temperature and pressure during recrystallization

grade of ore   percentage of the required metal or compound in the mined rock

granite   coarse-grained igneous rock with a high proportion of silica

greenhouse effect   warming of the Earth's surface by the thermal insulation provided by atmospheric gases

GNP   gross national product – the value of the goods and services produced by a national economy, from both domestic and overseas sources

groundwater   water held in the pores of subsurface rocks and soils

gypsum   hydrated calcium sulphate, a natural evaporite mineral

GWP   global warming potential of a gas over a particular future period, a function of its molecular structure, atmospheric concentration and lifetime

heat mining extraction of geothermal heat at a rate faster than it can be replenished by natural processes

heavy oil natural petroleum fraction with a viscosity too high to allow extraction by conventional pumping

holistic science investigation of systems as a whole rather than, as in reductionist science, in their separate parts

hot dry rock (HDR) system method of extracting geothermal energy by circulating water through otherwise dry rocks at depth

hydrocarbon compound made of hydrogen and carbon

hydrophone receiver for sound energy in a seismic reflection survey

hydrosphere the water masses of the Earth's surface – oceans, rivers, lakes, ice sheets, etc. – together with groundwater and atmospheric water

hydrothermal concerning the circulation of hot water in the subsurface

hyperthermal system source of geothermal energy from steam and hot water in areas of abnormally high crustal heat flow, usually due to volcanism or intrusion

igneous rock rock crystallized from the molten state

industrial mineral mineral valued for its own properties rather than those of its constituent metal

intangible resources assets such as scenery or biological diversity to which it is difficult to give a monetary value

intergenerational equity principle that the opportunities of future generations should not be adversely affected by the activities of the present generation

inter-regional equity principle that the lives of people in one region of the Earth should not be adversely affected by the activities of people in another region – also termed intragenerational equity

intrusion introduction of igneous rock into surrounding "country rock" by displacement or absorption

karst topography and landscape produced in limestone areas, by solution and related subsidence

kerogen waxy hydrocarbon produced from buried organic matter, which itself breaks down into oil or gas on further burial

lahar a mudflow composed of volcanic ash and water

lamination planar structure in sedimentary rocks spaced closer than 1 cm

landfill method of waste disposal by dumping in old quarries or natural topographic hollows

landslide general term for all types of downslope mass movement on land, but more specifically those in which displacement is concentrated along a basal shear plane

lava flow downslope stream of molten rock over the Earth's surface

leaching removal of elements or compounds from rocks or soils by solution

levee raised bank alongside a river channel that acts to constrain water during floods – may be natural or engineered

limestone sedimentary rock made mostly of calcium carbonate minerals

liquefaction transformation of unconsolidated granular sediment from a solid to a liquid state, for instance by seismic shaking

lithosphere relatively rigid outer layer of the Earth, about 100 km thick

magma partially or completely molten rock

mantle hot spot zone of anomalously hot upwelling mantle material

mass movement downslope movement of material under gravity

maturation action of temperature and time on buried organic matter to yield hydrocarbons

metamorphic rock rock formed by subjecting igneous or sedimentary rocks to temperatures or pressures different from those at their original formation

metasediment metamorphosed sedimentary rock

metasomatism metamorphism involving bulk addition or removal of chemical components other than water

methane hydrate a solidified mixture of methane gas in ice, also termed clathrate

migration movement of oil or gas through rock pores and fractures

monomineralic containing only one type of mineral

moraine landform made of glacial till

mudstone sedimentary rock dominated by grains less than 0.0625 mm in diameter

OECD Organization for Economic Cooperation and Development – comprises Australia, Austria, Belgium, Canada, Denmark, Finland, France, Germany, Greece, Iceland, Irish Republic, Italy, Japan, Netherlands, New Zealand, Norway, Portugal, Spain, Sweden, Switzerland, Turkey, UK and USA

oil-prone source hydrocarbon source rock with organic components that produce more oil than gas on burial

oil shale shale containing more than about 12% organic carbon

oolite or **oolitic limestone** sedimentary rock made of small spherical calcareous grains, formed by accretion of carbonate mud

ore rock containing a high enough proportion of valuable metal to make it economic to extract

orogeny episode of continental deformation involving contraction and uplift

peat unconsolidated deposit rich in plant matter

pegmatite   very coarsely crystalline igneous rock

periglacial   around the margins of an icesheet

permafrost   condition where ground remains frozen at depth throughout the year

permeability   capability of a rock or soil to transmit fluid through interconnected pores or fractures

photosynthesis   conversion by plants of carbon dioxide and water into carbohydrates and oxygen

piezometric head   pressure in the pores of a rock measured as an equivalent height of water

placer   sediment with ore minerals concentrated due to their high density or resistance to weathering

pore   gap between sedimentary grains

porosity   proportion of pore volume to the total volume of a rock

radiogenic   produced by decay of radioactive elements

rainout   transfer of atmospheric particles or aerosols to the ground surface by rain

recharge   replenishment of an aquifer by rainfall and subsurface infiltration

reductionist science   investigation of systems in their separate parts rather than, as in holistic science, as a whole entity

reef   specifically, a shallow marine rock mass bound by carbonate-depositing organisms, but often taken to include also any fringing units of carbonate debris

renewable resource   resource that can be replenished or reconditioned by natural processes

reserves   that part of the resources of a commodity that is economically expoitable now

reservoir   storage site, specifically a porous and permeable rock that contains oil and gas

residual deposit   rock that remains after low-temperature alteration, solution and transport of some of its original components

resource   the available stock of any useful commodity, specifically the amount that is economically available now or in the future

resource base   total stock of a commodity, including that which may never be economic to exploit

rheology   relationship of the shape change in a deforming rock to the imposed stresses

rhyolite   fine-grained igneous rock with a high proportion of silica

rifting   extension of the Earth's crust by faulting

risk   chance of an individual being affected, for instance, by a natural hazard – depends on both the level of hazard and the vulnerability of the region and individual

sag   thermal subsidence of a sedimentary basin

salt   usually refers to halite, less commonly to other evaporite minerals

sandstone   sedimentary rock composed mainly of grains between 0.0625 mm and 0.2 mm in diameter

schist   foliated metamorphic rock with grains visible to the naked eye

secondary enrichment   concentration of primary ore, often by solution and reprecipitation

seal or **cap rock**   the impermeable layer preventing the upward escape of oil or gas from a reservoir

sedimentary basin   site of sediment accumulation persisting through time and over an area of at least tens of square kilometres

sedimentary rock   rock formed by low temperature processes at the Earth's surface

seiche   large wave surge in an estuary or lake

seismic energy   energy radiating from the focus of an earthquake, most commonly generated by slip across faults

seismic section   a cross-section compiled from subsurface reflections of sound energy

seismometer   instrument for recording earthquake waves

shale   fine-grained sedimentary rock with a fissility

shear plane   a plane within which displacement is concentrated

shear strength   stress at which a rock fails under a force exerted parallel to its surface

silica sand   quartz sand of sufficient purity to be used for making glass

sinkhole   cylindical depression caused by ground subsidence above a zone of subsurface solution

slate   metamorphic rock with cleavage surfaces defined by flaky mineral grains too small to be visible with the naked eye or to give a sheen to the surfaces

slide   downslope mass movement concentrated on a basal shear plane

slump   a slide in which the basal shear plane is markedly concave upwards

smelt   to fuse or melt ore in order to extract metal

solid geology   rocks below the superficial cover of Quaternary sediments

source rock   organic-rich rock that can yield oil or gas on heating and burial

strain   change in shape of a rock body during deformation

stress   amount of force per unit area of a rock body exerted during deformation

sustainable resource   commodity that is used only at a rate at which natural processes can replenish and recondition it

syncline   fold with the youngest rocks in its core

syngenetic mineralization   formation of a mineral deposit at the same time as associated rocks

tar sand   sand or sandstone with bitumen or heavy oil in its pores

tectonic   pertaining to deformation of rocks

Tees-Exe Line   geographic line between the mouth of the River Tees in NE England and the mouth of the River Exe in SW England, roughly separating upland and lowland Britain

tensile strength   stress required to break a rock during extension

tephra   ejected volcanic debris

thermogenic   formed by heating e.g. gas from kerogen

threshold   a level marking a change in state in an environmental system

till or **boulder clay**   glacial deposit with coarse material supported in a fine-grained matrix

toxic   harmful above a certain concentration

transpiration   emission of water vapour from plants

trap   a rock geometry that concentrates upward migrating oil and gas

tsunami   large marine wave usually triggered by a submarine earthquake or eruption

turbidite   sedimentary product of a turbidity current – a turbulent downslope flow of sediment and water

unconfined aquifer   an aquifer with its top open to the Earth's surface rather than overlain by impermeable rocks

unconformity   surface representing a time break in rock accumulation – may be **angular** if underlying rocks have been tilted during this time

unconventional petroleum   oil and gas in deep or otherwise poorly accessible reservoirs

unimodal   having two maxima on a distribution curve of a natural parameter such as grain size or mineral abundance

upstream   towards the earlier stages of an industrial or economic process

Variscan   deformation event or orogeny in Late Carboniferous and early Permian time

vein   thin planar zone of precipitated minerals or intruded rock

volatile organic compounds (VOC)   a range of gaseous pollutants derived by evaporation of fuels and solvents

vulnerability   the extent to which different populations and their natural or built environments are susceptible to particular level of natural hazard

wall rock   host rock to a hydrothermal vein or igneous intrusion

water table   surface below which rocks are saturated by groundwater

weathering   breakdown or solution of rock *in situ* by mechanical, chemical or biological processes

wet gas   natural gas which yields liquid condensates when cooled

REFERENCES

Listed here are the sources of illustrations cited on page viii. Many of these contain the primary data from which graphs and figures have been constructed. These will be valuable for more detailed analysis or study by readers.

Anderson, J. G. C. & C. F. Trigg 1976. *Case-histories in engineering geology.* London: Elek Science.

Avery, B.W. 1990. *Soils of the British Isles.* Wallingford, Berks.: CAB International.

Bailey, M. C., J. J. Bowman, C. O'Donnell 1986. *Air quality in Ireland: the present position.* Dublin: An Foras Forbatha.

Bookout, J. F. 1989. Two centuries of fossil fuel energy. *Episodes* 12, 257–62.

Bottomley, A. 1992. Graduate geological scientists' survey 1990. *Geoscientist* 2(2), 17–19.

Boulton, G. S., A. S. Jones, K. M. Clayton, M. J. Kenning 1977. A British ice-sheet model and patterns of glacial erosion and deposition in Britain. In *British Quaternary Studies: recent advances,* F. W. Shotton (ed.), 231–46. Oxford: Clarendon.

Brassington, F. C, and K. R. Rushton 1987. A rising water table in central Liverpool. *Quarterly Journal of Engineering Geology* 20, 151–8.

Brennand, T. P., B. Van Hoorn, K. H. James 1990. Historical review of North Sea exploration. In *Introduction to the petroleum geology of the North Sea,* K. W. Glennie (ed.), 1–33. Oxford: Blackwell Scientific.

British Gas 1990. *Oil and gas fields and the national gas transmission system.* Map BG1014. London: British Gas.

British Geological Survey 1979. *Merthyr Tydfil. 1:50,000 geological map.* Southampton: Ordnance Survey.

British Geological Survey 1984. *Aberystwyth. 1:50,000 geological map.* Southampton: Ordnance Survey.

British Geological Survey 1992. *World mineral statistics 1986–1990.* Keyworth, Nottingham: British Geological Survey.

British Geological Survey 1993. *United Kingdom minerals yearbook 1992.* Keyworth, Nottingham: British Geological Survey.

British Petroleum 1993. *BP statistical review of world energy 1993.* London: British Petroleum Company p.l.c.

Brookes, A., K. J. Gregory, F. H. Dawson 1983. An assessment of river channelization in England and Wales. *Science of the Total Environment.* 27, 97–112.

Brown, G. C. & E. Skipsey 1986. *Energy resources: geology, supply and demand.* Milton Keynes, Bucks.: Open University Press.

Carr, D. D. & N. Herz (eds) 1989. *Concise encyclopedia of mineral resources.* Oxford: Pergamon.

Chicken, J. C. 1975. *Hazard control policy in Britain.* London: Pergamon.

Cooper, A. H. 1986. Subsidence and foundering of strata caused by the dissolution of Permian gypsum in the Ripon and Bedale areas, North Yorkshire. *Special Publication of the Geological Society, London* 22, 127–39.

Cornford, C. 1990. Source rocks and hydrocarbons of the North Sea. In *Introduction to the petroleum geology of the North Sea,* K. W. Glennie (ed.), 294–361. Oxford: Blackwell Scientific.

Cripps, J. C. & C. C. Hird 1992. A guide to the landslide at Mam Tor. *Geoscientist* 2(3), 22–7

Cruickshank, J. G. & D. N. Wilcock (eds) 1982. *Northern Ireland: environment and natural resources.* Belfast: Queen's University and New University of Ulster.

Darby, H. C. 1969. *The draining of the fens,* reprint. Cambridge: Cambridge University Press.

Darmstadter, J. 1971. *Energy in the world economy.* Baltimore: Johns Hopkins University Press.

Department of the Environment 1984. *Survey of derelict land in England 1982.* London: HMSO.

Department of the Environment 1986. Pollution paper no. 26: *Nitrate in water.* London: HMSO.

Department of the Environment 1991. *Survey of derelict land in England 1988.* London: HMSO.

Department of the Environment 1992. *The UK environment.* London: HMSO.

Downing, R.A. & D. A. Gray 1986. *Geothermal energy: the potential in the United Kingdom.* London: HMSO.

Dunning, F. W., I. F. Mercer, M. P. Owen, R. H. Roberts, J. L. M. Lambert 1978. *Britain before man.* London: HMSO.

Dury, G. H. 1973. *The British Isles,* 5th edn. London:

REFERENCES

Heinemann.

Energy Technology Support Unit 1985. Prospects for exploitation of the renewable energy technologies in the UK. *Report of the Department of Energy* R30.

Energy Technology Support Unit 1987. Prospects for UK renewable energy technologies. Review: *Quarterly journal of renewable energy* 1, 4–5.

Energy Technology Support Unit 1989. *Renewable energy in Britain*. London: Department of Energy.

Eyre, W. A. 1973. The revetment of rock slopes in the Clevedon Hills for the M5 motorway. *Quarterly Journal of Engineering Geology* 6, 223–9.

Fielding, C. R. 1984. A coal depositional model for the Durham Coal Measures of NE England. *Journal of the Geological Society of London* 141, 919–31.

Flood, M. 1986. *Energy without end: the case for renewable energy*. London: Friends of the Earth.

Francis, E. H. 1979. British coalfields. *Science Progress* 66, 1–23.

Gallois, R. W. 1978. A pilot study of oil shale occurrences in the Kimmeridge Clay. *Reports of the Institute of Geological Sciences* 78/13, 1–24.

Geological Museum 1988. *Britain's offshore oil and gas*. London: British Museum (Natural History).

Gillmor, D. A. (ed.) 1979 *Irish resources and land use*. Dublin: Institute of Public Administration.

Goudie, A. 1990. *The landforms of England and Wales*. Oxford: Basil Blackwell.

Gowring, G. I. B. & A. Hardie 1968. Severn Bridge: foundations and substructure. *Proceedings of the Institution of Civil Engineers* 41, 49–68

Guion, P. D. & C. R. Fielding 1988. Westphalian A + B sedimentation in the Pennine Basin, UK. In *Sedimentation in a synorogenic basin complex*, B. M. Besly & G. Kelling (eds), 153–77. Glasgow: Blackie.

Houghton, J. T., G. J. Jenkins, J. J. Ephraums (eds) 1990. *Climate change: the IPCC scientific assessment*. Cambridge: Cambridge University Press.

Houghton, J. T., B. A. Callander, S. K. Varney (eds) 1992. *Climate change 1992: the supplementary report to the IPCC scientific assessment*. Cambridge: Cambridge University Press.

Johnston, R. J. & J. C. Doornkamp 1982. *The changing geography of the United Kingdom*. London: Methuen.

Lumsden, G. I. (ed.) 1992. *Geology and the environment in western Europe*. Oxford: Oxford University Press.

L'vovich, M.I. 1979. *World water resources and their future*. translation from 1974 edition in Russian, R. L. Nace (ed.). Washington, DC: American Geophysical Union.

Macdonald, G. J. 1990. The future of methane as an energy resource. *Annual Review of Energy* 15, 53–83.

Masters, C. D., E. D. Attanasi, W. D. Dietzman, R. F.

Meyer, R. W. Meyer, R. W. Mitchell, D. H. Root 1987. World Resources of crude oil, natural gas, natural bitumen, and shale oil. *Proceedings of the 12th World Petroleum Congress* 5, 3–27.

Masters, C. D., D. H. Root, E. D. Attanasi 1990. World resources of crude oil, natural gas, natural bitumen, and shale oil. *Annual Review of Energy* 15, 23–51.

Nisbet, E. G. 1991. *Leaving Eden: to protect and manage the Earth*. Cambridge: Cambridge University Press.

Parsley, A. J. 1990. North Sea hydrocarbon plays. In *Introduction to the petroleum geology of the North Sea*, K. W. Glennie (ed.), 362–88. Oxford: Blackwell Scientific.

Penoyre, J. & J. Penoyre 1978. *Houses in the landscape*. London: Faber & Faber.

Perry, A. H. 1981. *Environmental hazards in the British Isles*. London: Allen & Unwin.

Porter, E. 1978. *Water management in England and Wales*. Cambridge: Cambridge University Press.

Raynaud, D., J. Chappellaz, J. M. Barnola, Y. S. Korotkevich, C. Lorius 1988. Climatic and CH_4 cycle implications of glacial-interglacial CH_4 change in the Vostok ice core. *Nature* 333, 655–7.

Robinson, D. A. & R. B. G. Williams 1984. *Classic landforms of the Weald*. Sheffield: Geographical Association.

Shell 1991. Energy in profile. *Shell Briefing Service* 91/3, 1–13. London: Shell.

Skinner, B. J. 1987. Supplies of geochemically scarce metals. In *Resources and world development*, D. J. McLaren & B. J. Skinner (eds), 305–25. Chichester, England: John Wiley.

Smith, K. 1992. *Environmental hazards*. London: Routledge.

Steers, J. A. 1981. *Coastal features of England and Wales*. Cambridge: Oleander Press.

Taylor, J. C. M. 1990. Upper Permian–Zechstein. In *Introduction to the petroleum geology of the North Sea*, K. W. Glennie (ed.), 1–33. Oxford: Blackwell Scientific.

Thompson, S. A. 1982. Trends and developments in global natural disasters, 1947 to 1981. *Working papers, University of Colorado Institute of Behavioural Science* 45.

Tooley, M. J. 1993. Long term changes in eustatic sea level. In *Climate and sea level change: observations, projections and implications*, R. A. Warwick, E. M. Barrow & T. M. L. Wigley (eds), 81–107.

United Kingdom Nirex Limited 1987. *The way forward: a discussion document*. Harwell: UK Nirex.

United Nations 1992. *UN Statistical Yearbook*. 37th issue (for 1988/89). New York: United Nations Publications.

Wallwork, K. L. 1956. Subsidence in the Mid-Cheshire industrial area. *Geographical Journal* 122, 40–53.

Wallwork, K. L. 1974. *Derelict land*. Newton Abbott: David & Charles.

Waltham, A. C. 1989. *Ground subsidence*. Glasgow: Blackie.

Watson, J. W. & J. B. Sissons 1964. *The British Isles: a sys-*

tematic geography. London: Nelson.

Weisser, J. D., M. Dalheimer, D. Kohler, B. Sarbas 1987. Sources and production of widely used metals: an assessment of world reserves and resources. In *Resources and world development,* D. J. McLaren & B. J. Skinner (eds), 245–304. Chichester, England: John Wiley.

White, D. A. 1987. Conventional oil and gas resources. In *Resources and world development,* D. J. McLaren & B. J. Skinner (eds), 113–28. Chichester, England: John Wiley.

Williams, G. M., C. A. M. Ross, A. Stuart, S. P. Hitchman, L. S. Alexander 1984. Controls on contaminant migration at the Villa Farm Lagoons. *Quarterly Journal of Engineering Geology* 17, 39–55.

World Energy Conference 1986. *Survey of energy resources*. London: World Energy Conference.

World Resources Institute 1992. *World Resources 1992–3*. Oxford: Oxford University Press.

INDEX

spoil heaps 41
springs 46, 47
steam cap 98
steel 18
stone 15, 16–17
  dimension 52, 55, 56, 57
stopes, in mines 111
stratosphere 25
strip mining 112
stripping ratio 112
strips, foundation 117
sub-base, road 54
sub-bituminous coal 73, 77
subsidence
  ground 2, 108, 109, 110–11
  Quaternary 13
  tectonic 30, 72
subsidence bowl 110
Sudbury, Canada 115
Suffolk 111
sulphur dioxide 73, 76, 91, 102, 103,
  106, 130, 134, 135
superquarries 58, 114
surface water 44, 47, 50
surface waves 24
sustainability, of resources 34, 35,
  105
swelling, of ground 117
switching, of delta lobes 71, 72
syngenetic minerals 63, 66, 67
systems science 105

Taf Fechan, Afon (river) 121
talc 66, 67, 68
tar sands 90
Tees estuary 19, 21
Tees-Exe line 9, 10, 14, 48, 56, 57
tensile strength, of rock 116
terraces, river 57
textiles, manufacture of 19
Thames estuary 19
Thames, River 113
thatch, as roofing 16, 17
thematic maps 41
thermal conductivity 99
thermal reactors 96
thermogenic gas 79
threshold, of damage 23
throughstones, in walls 17
thunderstorms 23
tidal energy 100, 101
tight-reservoired gas 91

tiles, as roofing 16, 52
till 13, 57
timber, in buildings 16
tin 20, 61, 64, 66, 67, 68, 69, *see
  also* cassiterite
titanium 61, 68, 69 *see also* ilmenite
toppling, in rock failure 118
tornadoes 23
toxicity, geomedical 31
Trail, Canada 115
transpiration, in carbon cycle 44
trap, petroleum 78, 80, 81, 82, 84,
  85
trough, glacial 12
tsunamis 23, 25
tungsten 61, 66, 67, 68, 69, *see also*
  wolframite
tunnels 118, 119
turbidites 70, 71
turf, in buildings 17
Tyne estuary 19

unconfined aquifer 46–7
unconformities 8, 80, 81, 85, 97
unconventional petroleum 78, 90–91
unimodal grade curve 37
unsaturated zone 44, 46
up-dip face, of excavation 118
uplands 10, 11, 14, 40
uplift
  Quaternary 13
  Palaeogene 9, 10
upstream, in industrial process 106
uraninite 97
uranium 62, 66, 96–7, 115
urbanization 32, 120, 124
use
  construction materials 58
  minerals 60, 61
  petroleum 91
utilitarian ethic 145

valley bog 72
vanadium 61, 68, 69
Variscan Front 75
Variscan granites 64, 65, 66, 99
Variscan orogeny 6, 7, 8, 9, 64, 65,
  67, 73
Variscides 8
veins, mineral 61, 62, 63, 64, 97
vernacular architecture 16–17
Viking Graben 84, 89

Villa Farm, Coventry 128
vitrinite 73
volatile organic compounds (VOC) 135
volatiles, in coal 73
volcanoes 2, 23, 25, 138, 140
vulnerability, to hazards 23, 32

walls, dry stone 15, 17
Wash, Tthe 13
wastes 105
  agricultural 126, 127, 131
  atmospheric 2, 130
  liquid 2, 126, 127, 128, 129
  radioactive 132
  solid 2, 126, 127, 129
water 2, 44–51
  demand for 50
  subsurface 2
  use of 45, 51
  world resources 51
water table 26, 39, 44, 46–7, 49,
  118–19, 124, 128
water table well 47
Watson, J. 143
wattle and daub 16
wave energy 100, 101
waves, seismic 24
Weald 11, 57
Weald Clay 56, 57
wearing course, of road 54
weathering 27, 39
  biological 39
  chemical 39
  mechanical 39
wells 47, 108, 109, 111
Welsh Borderland 11, 29, 57
Wessex Basin 99
wet gas 79
wetlands 40
wind energy
wireline logs 82
wolframite 61, 64, *see also* tungsten
woodland 40, 42
Worcester Basin 99

yellowcake, uranium 96
Yorkshire 17, 30

Zechstein evaporites 65, 84, 85, 86
zinc 31, 61, 64, 66, 67, 68, 69, *see
  also* sphalerite